Adobe Illustrator

LISA FRIDSMA

 Peachpit Press

Visual QuickStart Guide
Adobe Illustrator
Lisa Fridsma

Peachpit Press
www.peachpit.com

Copyright © 2022 by Pearson Education, Inc. or its affiliates. All Rights Reserved.
San Francisco, CA
Peachpit Press is an imprint of Pearson Education, Inc.

To report errors, please send a note to errata@peachpit.com

Executive Editor: Laura Norman
Development Editor: Robyn G. Thomas
Senior Production Editor: Tracey Croom
Technical Editor: Monika Gause
Copy Editor: Robyn Thomas
Proofreader: Kelly Anton
Compositor: Lisa Fridsma
Indexer: Lisa Fridsma
Cover Design: RHDG / Riezebos Holzbaur Design Group, Peachpit Press
Interior Design: Peachpit Press with Danielle Foster
Logo Design: MINE™ www.minesf.com

ISBN-13: 978-0-13-759774-1
ISBN-10: 0-13-759774-6

1 2022

Dedication

To my family, both by blood and by choice—
Your support and love is staggering.

Special Thanks to:

Laura Norman for giving me this amazing opportunity and providing support throughout the journey.

Robyn Thomas for managing with a steady hand and keen eyes throughout the process.

Von Glitschka for lending his expertise and voice in producing the videos.

Tracey Croom for being my production editing hero.

Monika Gause for being my technical editing hero.

Kelly Anton for being my proofreading hero.

Pearson's Commitment to Diversity, Equity, and Inclusion

Pearson is dedicated to creating bias-free content that reflects the diversity of all learners. We embrace the many dimensions of diversity, including but not limited to race, ethnicity, gender, socioeconomic status, ability, age, sexual orientation, and religious or political beliefs.

Education is a powerful force for equity and change in our world. It has the potential to deliver opportunities that improve lives and enable economic mobility. As we work with authors to create content for every product and service, we acknowledge our responsibility to demonstrate inclusivity and incorporate diverse scholarship so that everyone can achieve their potential through learning. As the world's leading learning company, we have a duty to help drive change and live up to our purpose to help more people create a better life for themselves and to create a better world.

Our ambition is to purposefully contribute to a world where:

- Everyone has an equitable and lifelong opportunity to succeed through learning.
- Our educational products and services are inclusive and represent the rich diversity of learners.
- Our educational content accurately reflects the histories and experiences of the learners we serve.
- Our educational content prompts deeper discussions with learners and motivates them to expand their own learning (and worldview).

While we work hard to present unbiased content, we want to hear from you about any concerns or needs with this Pearson product so that we can investigate and address them.

- Please contact us with concerns about any potential bias at www.pearson.com/report-bias.html.

Contents at a Glance

Table of Contents

LIST OF VIDEOS

LIST OF VIDEOS

List of Videos continues

LIST OF VIDEOS

Introduction

Welcome to Adobe Illustrator, the industry-standard vector graphics application. Illustrator's tools and features provide artists, designers, and illustrators endless possibilities for creating logos, icons, drawings, typography, and complex illustrations for any printed or digital purpose.

Here, we'll share some thoughts to help you get the most from this book. Then you can be on your way to perfecting using the simple, fun, and sophisticated graphics tools in Illustrator.

How to Use this Book

This Visual QuickStart Guide, like others in the series, is a task-based reference. Each chapter focuses on a specific area of the application and presents it in a series of concise, illustrated steps.

This book is meant to be a reference work, and although it's not expected that you'll read through it in sequence from front to back, we've made an attempt to order the chapters in a logical fashion.

The first chapter covers an overview of the Illustrator application interface and walks you through creating a new document. The next two chapters steps you through customizing the application and documentss to best suit your needs.

From there you dive into the rich features Illustrator provides for creating colorful and visually interesting designs. The last chapter teaches you a variety of techniques for saving and exporting your artwork.

This book is suitable for the beginner just starting in vector drawing and design, as well as hobbyists, intermediate-level designers, and illustrators.

Sharing Space with Windows and macOS

Illustrator is almost exactly the same on Windows and macOS, which is why this book covers both platforms.

In the few places where a feature is found in one environment but not the other, or if the steps are different for each, we make it clear which version is being discussed.

You'll also see that the screenshots are macOS—but despite a few cosmetic differences such as title bars and menu bars, everything pretty much tracks the same within the user interface itself.

We also include keyboard shortcuts for both platforms as an appendix at the end of the book. Keyboard shortcuts are faster methods of accessing commands compared to choosing items from menus. They are also used and described often in the supplemental video tutorials.

Speaking of...this video icon makes it easy to find the videos to watch in the Web Edition:

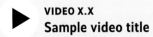

VIDEO X.X
Sample video title

TIP To better view the interface onscreen or in print, the screen captures in this book reflect the **Medium Light** interface rather than the default setting of **Dark**. You can modify interface settings in Preferences.

Online Content

Your purchase of this Visual QuickStart Guide includes a free online edition of the book, which contains the videos and is accessed from your Account page on www.peachpit.com.

Web Edition

The Web Edition is an online interactive version of the book, providing an enhanced learning experience. You can access it from any device with a connection to the internet, and it contains the following:

- The complete text of the book
- Over 40 instructional videos keyed to the text

Accessing the Web Edition

TIP If you encounter problems registering your product or accessing the Web Edition, go to www.peachpit.com/support for assistance.

You must register your purchase on peachpit.com in order to access the online content:

1. Go to **www.peachpit.com/ illustratorvqs2022**.
2. Sign in or create a new account.
3. Click **Submit**.
4. Answer the question as proof of purchase.
5. Access the **Web Edition** from the **Digital Purchases** tab on your Account page.
6. Click the **Launch** link to access the product.

If you purchased a digital product directly from peachpit.com, your product will already be registered. However, you still need to follow the registration steps and answer the proof-of-purchase question to access the Web Edition.

1

The Basics

Adobe Illustrator is the industry standard for creating vector-based illustrations.

While Illustrator was originally used for simple elements such as logos and icons, its robust features and user-friendly interface have evolved to provide the ability to create rich and complex designs.

In This Chapter

The Illustrator Interface

The Illustrator interface is designed for an intuitive and user-friendly experience.

Home screen overview

Starting Illustrator directly (without opening an Illustrator file) takes you to the **Home** screen (**Figure 1.1**), which provides a variety of tools and resources for getting you started.

TIP To learn about starting Illustrator, see the **"Launching Illustrator" section in this chapter.**

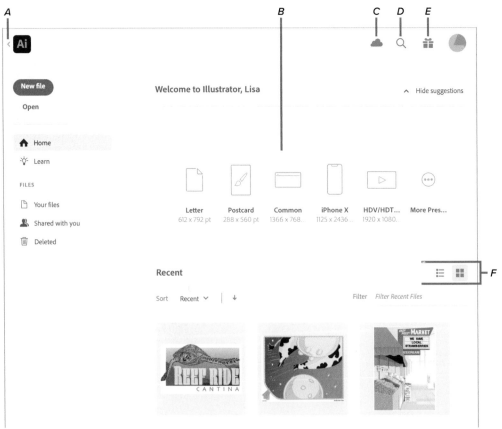

FIGURE 1.1
A. Back to Application Frame button **B.** Presets for new documents **C.** Cloud storage status **D.** Search button
E. What's New button **F.** Display Recent Files by List or Preview toggles

Application frame overview

The Illustrator application frame (**Figure 1.2**) is a customizable interface that lets you easily access and configure a variety of tools and panels to help create and modify your artwork.

The appearance of the application frame depends on which workspace is active. This book focuses primarily on the **Essentials Classic** workspace.

TIP To learn more about workspaces, see the "Customizing Workspaces" section in Chapter 2.

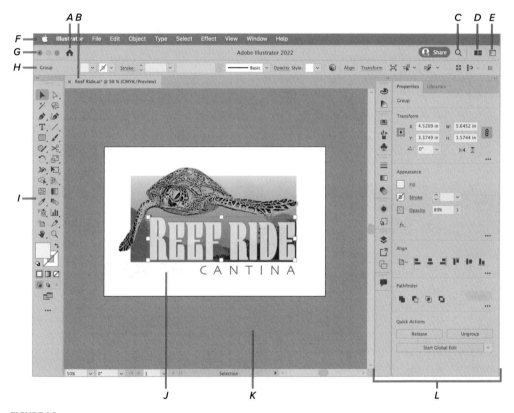

FIGURE 1.2
A. Return to Home Screen button **B.** Document window tab **C.** Learn button **D.** Arrange Documents button
E. Workspace menu **F.** Menu bar **G.** Application bar **H.** Control panel **I.** Toolbar **J.** Artboard **K.** Pasteboard
L. Panels section

Launching Illustrator

Illustrator can be launched either directly or by opening an Illustrator (.ai) file.

Start the application

Do any of the following:

- Click the **Adobe Illustrator 2022** icon on your desktop dock (macOS) or Start screen (Windows).

- Locate the **Adobe Illustrator 2022** application folder and double-click the application icon (**Figure 1.3**).

- Double-click an **Illustrator (.ai)** file (**Figure 1.4**).

TIP Depending on your system, file icons may display as either the application type or a thumbnail of the document.

FIGURE 1.3 Double-clicking the application in the Adobe Illustrator 2022 folder to launch it

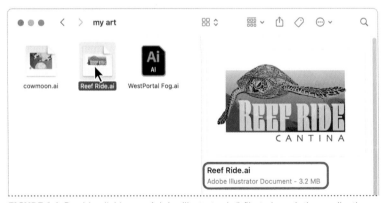

FIGURE 1.4 Double-clicking an Adobe Illustrator (.ai) file to launch the application

Learning about Illustrator

The **Discover** panel (**Figure 1.5**) is a new feature that provides quick integrated access to a variety of learning resources.

Open the Discover panel from the Home screen

Do the following:

- Click **Learn** in the left panel and then select a tutorial.

- Click the **What's New** button (**E** in Figure 1.1) to display information about new features in the application and how to use them.

Open the Discover panel from the application frame

Do any of the following:

- Choose **Help** > **Illustrator Help** to display the main menu.

- Choose **Help** > **Tutorials** to display the in-app **Tutorials** section.

- Choose **Help** > **What's New** to display information about new features in the application and how to use them.

- Click the **Search** button (**C** in Figure 1.2).

TIP Clicking the Home button in the upper right of the Tutorials and What's New sections takes you to the main menu.

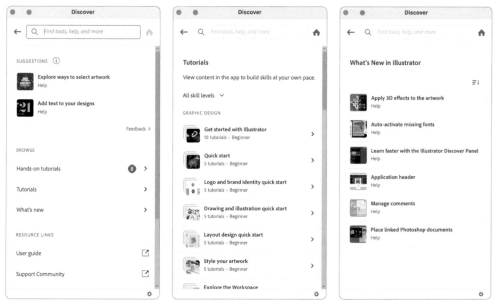

FIGURE 1.5 The Discovery panel displaying the main menu, Tutorials, and What's New sections

Opening an Existing File

There are numerous ways to open a file.

Open a file from the Home screen

Do any of the following:

- Click the **Open** button, and then navigate to select the file (**Figure 1.6**).

- Click one of the options under **Files** and select the file.

- Click a file listed under the **Recent** section.

Open a file from the application frame

Do any of the following:

- Choose **File** > **Open** and navigate to the desired file.

- Choose **File** > **Open Recent Files** and select a recently opened file from the context menu.

- Choose **File** > **Browse in Bridge** to launch Adobe Bridge and select a file using that application (**Figure 1.7**).

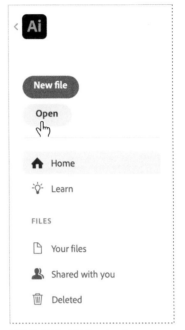

FIGURE 1.6 Clicking the Open button in the Home screen

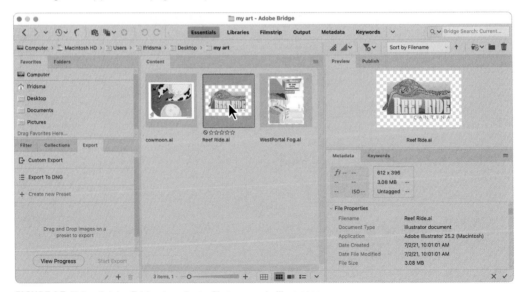

FIGURE 1.7 Using Adobe Bridge to select a file to open in Illustrator

Creating a New File

New files can be created from the **Home** screen or within the application.

Create a file from the Home screen

Do any of the following:

- Click **New File** to open the New Document dialog box.
- Click a preset to create and automatically open a new document in the application frame (**Figure 1.8**).
- Click **More Presets** to open the New Document dialog box.

Create a file from the application frame

Do any of the following:

- Choose **File** > **New** to open the New Document dialog box.
- Choose **File** > **New from Template** and select the appropriate template file.

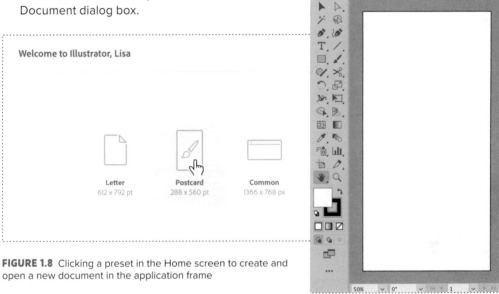

FIGURE 1.8 Clicking a preset in the Home screen to create and open a new document in the application frame

TIP Document settings can be changed after the file is created in the Document Setup dialog box (File > Document Setup).

Use presets in the New Document dialog box to create a document

Do the following:

1. At the top of the dialog box, select a category tab (**Figure 1.9**).

2. Select one of the presets below the corresponding category.

3. Click **Create**.

TIP The Saved category tab contains templates downloaded from Adobe Stock.

TIP The Templates section contains downloadable templates from Adobe Stock.

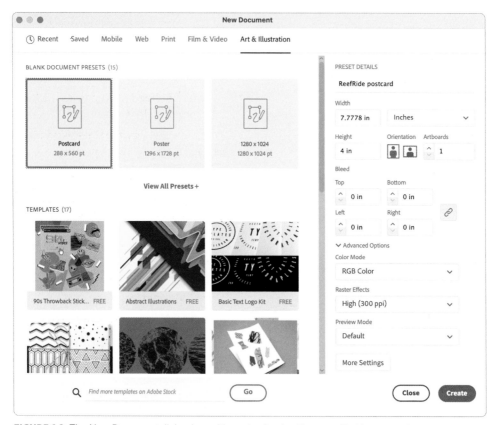

FIGURE 1.9 The New Document dialog box with customized settings applied to a preset

What is a bleed?

Bleeds are elements that extend beyond the printable boundaries of a document. The portion of the element bleeding beyond the boundary will not be printed.

In Illustrator, the edge of the artboard defines the bleed boundary. However, they can also be specified to reside outside the artboard (**Figure 1.10**).

FIGURE 1.10 The bleed positioned outside this artboard is indicated by the red border.

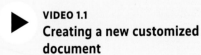

VIDEO 1.1
Creating a new customized document

Customize presets for a new document

On the right side of the **New Document** dialog box, do any of the following:

- Enter a **name** for the document.
- Use the **Width** and **Height** fields to specify the artboard size.
- Select the unit of measurement from the context menu.
- Select portrait or landscape for the **Orientation**.
- Increase or reduce the number of **Artboards**.
- Specify the **Bleed** positions for each edge of the artboard.
- Select the appropriate **Color Mode** for your project.

TIP By default, Illustrator automatically selects the appropriate color mode for printed (CMYK) or digital (RGB) outputs. To learn more about color modes, see Chapter 4.

- Customize the resolution of any **Raster Effects** applied to the document.
- Select the appropriate **Preview Mode** for viewing your artwork:

  Default displays all art as vector and in full color.

  Pixel simulates how the art will appear if rasterized.

  Overprint simulates how the art will appear when printed.

Working with Templates

Templates are useful when creating multiple documents that share similar components (dimensions, color modes, etc.).

Open an application-provided template file

Illustrator provides some industry-standard template files (.ait) with the application. Selecting a template opens a copy of the file as a blank, unsaved Illustrator document.

1. Choose **File** > **New from Template**.

2. In the dialog box, open the **Blank Templates** folder located under:

   Adobe Illustrator 2022 / Cool Extras / en_US / Templates

3. Select the appropriate file and then click **New** (**Figure 1.11**).

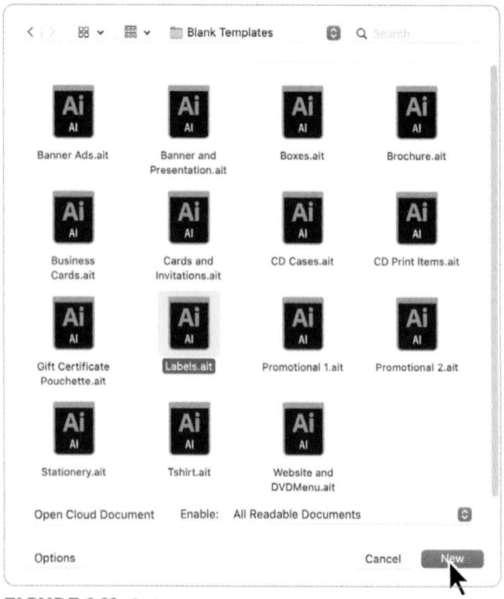

FIGURE 1.11 Selecting a template provided with the Illustrator application

Create a template from an existing file

When you create a template using an existing document, all the artwork and application settings are retained when you open a copy as an .ai file.

1. Customize the document so it displays as you want it to when opened.

2. Choose **File** > **Save as Template**.

3. Assign a name and location for the template file in the Save As dialog box.

4. Click **Save** to complete saving the file as a template.

Customizing your template file

These are some of the ways you can customize your document prior to saving it as a template:

- Setting the magnification level

- Displaying rulers, guides, or grids

- Including and excluding artwork

- Including and excluding swatches, brushes, symbols, etc.

- Using the Document Setup and/or Print Setup dialog boxes to set the desired options

Customizing
the Application

Adobe Illustrator provides vast customization capabilities to best suit your needs
when using the application.

You can organize tools and panels as
needed and save those settings for
future use.

In This Chapter

Accessing the Document Windows

Illustrator lets you have multiple files open simultaneously.

Work with nested documents

By default, files are nested in the document window.

- Activate a document by clicking its tab (**Figure 2.1**).
- Close a document by clicking the **X** icon on the tab.

Work with floating documents

Documents can also reside apart from the application frame. This can be useful if you are using multiple monitors.

- Float a nested document by dragging its tab from the application frame.
- Nest a floating document by dragging its title bar onto the application frame (**Figure 2.2**).

FIGURE 2.1 Clicking a nested tab activates the document

TIP You can determine how documents open using Preferences > User Interface. The default setting is Open Documents as Tabs.

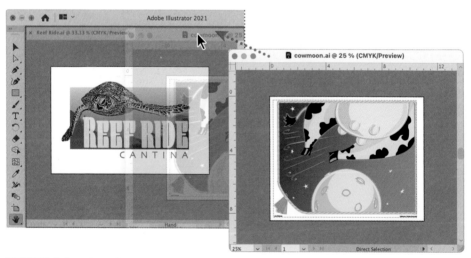

FIGURE 2.2 Dragging a floating document by its title bar to nest it in the application frame

Tile the document windows

To tile all open documents so they are visible in the application frame, do either of the following:

- Click the **Arrange Documents** button in the application bar and select a tiled option (**Figure 2.3**).
- Choose **Window** > **Arrange** > **Tile**.

Consolidate the document windows

To gather all open documents so they are nested and tabbed in the application frame, do either of the following:

- Click the **Arrange Documents** button in the application bar and click the **Consolidate All** icon (top left).
- Choose **Window** > **Arrange** > **Consolidate All Windows**.

FIGURE 2.3 Clicking the Arrange Documents button displays options for tiling and consolidating open windows

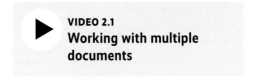

VIDEO 2.1
Working with multiple documents

Using the Toolbar

By default, the various tools associated with an assigned workspace reside in the toolbar, which is docked on the left side of the application frame.

Select a tool

Do either of the following:

- Click the tool in the toolbar.
- Press the keyboard shortcut for the tool.

TIP The keyboard shortcut for a tool is shown in parentheses after the tool name when you hover your mouse over it.

Show hidden tools

Similar tools are organized in groups and identifiable by a small triangle in the lower-right corner of the visible tool indicating hidden tools. The hidden tools are accessible by doing either of the following:

- Click+hover over the visible tool.
- Press **Alt/Option**+click to cycle through the individual hidden tools.

Reposition the toolbar

The toolbar can be undocked and moved by doing the following:

- Click+drag the title bar to the toolbar's desired location.

Float a tool group

Do the following:

- Click+drag the tool group tear-off tab.

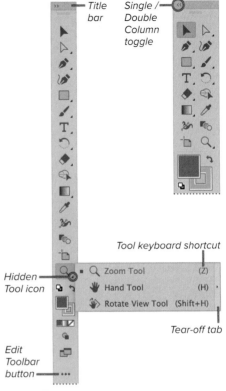

FIGURE 2.4 Toolbar customization features

View tools in double or single column

Do the following:

- At the top left of the toolbar, click the double arrows (**Figure 2.4**).

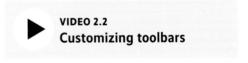

VIDEO 2.2
Customizing toolbars

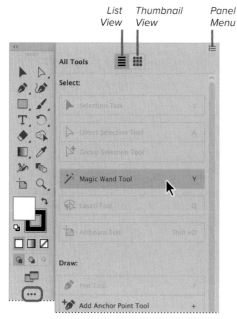

List View *Thumbnail View* *Panel Menu*

FIGURE 2.5 Selecting a tool from the All Tools drawer

FIGURE 2.6 Creating a new toolbar using the All Tools drawer panel menu

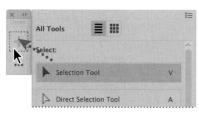

FIGURE 2.7 Adding a tool to a new toolbar

Access the All Tools drawer

The **All Tools** drawer (**Figure 2.5**) contains every tool provided by Illustrator. To access them, do the following:

1. Click the **Edit Toolbar** button.

2. Click a tool to select it.

TIP Tools can be selected and used without adding them to the toolbar.

Add a tool to the toolbar

Do the following (**Figure 2.6**):

1. Click the **Edit Toolbar** button to open the **All Tools** drawer.

2. Click+drag the tool onto the toolbar.

TIP Tools that appear dimmed in the All Tools drawer reside in the toolbar.

Delete a tool from the toolbar

Do the following:

1. Click the **Edit Toolbar** button to open the **All Tools** drawer.

2. Click+drag the tool away from the toolbar.

Create a new toolbar

Do the following:

1. Open the **New Toolbar** dialog box by doing either of the following:

   Choose **Windows** > **Tools** > **New Toolbar**.

   Click the **Edit Toolbar** button and then choose **New Toolbar** from the tool drawer panel menu.

2. In the New Toolbar dialog box, enter a **Name**, and then click **OK** (**Figure 2.7**).

   The new empty toolbar will appear floating (undocked) on your screen.

Working with Panels

The numerous panels included with Illustrator provide powerful tools for creating and modifying your artwork. Because there are so many, Illustrator lets you easily access and organize them as needed to best suit your work needs (Figure 2.8).

Open a closed panel

Do the following:

- Choose **Window** > *[panel name]*.

TIP Depending on your workspace configuration, panels may open as docked or floating, individually or within a group.

Open a collapsed panel

Collapsed panels display only their icons. To open one, do any of the following:

- Choose **Window** > *[panel name]*.
- Click the icon for the collapsed panel.
- Click the **Expand Panels** button to open all the collapsed panels in the dock (Figure 2.9).

Close a docked panel

Do the following:

1. Right-click the panel tab or icon.
2. Choose **Close** from the context menu.

Close a floating panel

Do the following:

- Click the panel **Close** button (Figure 2.10).

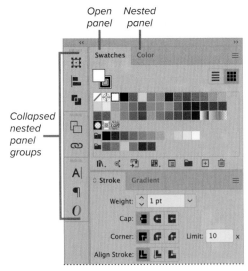

Open panel Nested panel

Collapsed nested panel groups

FIGURE 2.8 Collapsed and open panel groups

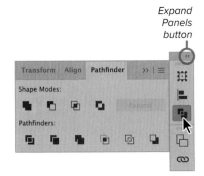

Expand Panels button

FIGURE 2.9 Expanding a collapsed panel

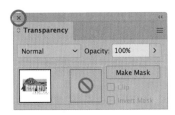

FIGURE 2.10 Close button on a floating panel

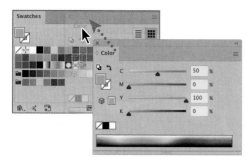

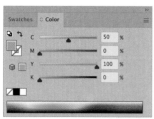

FIGURE 2.11 Nesting a floating panel to create a panel group

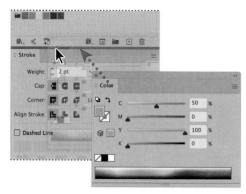

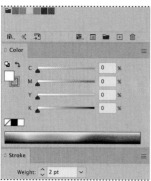

FIGURE 2.12 Docking a floating panel

Move a panel

Do the following:

- Click+drag the panel tab.

Move a panel group

Do the following:

- Click+drag the group title bar.

Nest a panel in a group

Do the following:

- Click+drag the panel tab onto the group (**Figure 2.11**).

TIP A blue boundary around the destination panel indicates you are creating a group.

Dock a panel

Do the following:

- Click+drag the panel tab above or below another docked panel (**Figure 2.12**).

TIP A blue horizontal highlight indicates you are docking the panel.

Dock a panel group

Do the following:

- Click+drag the panel title bar above or below another docked panel.

Maximize or minimize panels

Do the following:

- Double-click the panel title bar (**Figure 2.13**).

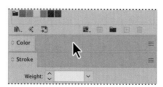

FIGURE 2.13 Color panel minimized after double-clicking the title bar

Working with the Properties Panel

The **Properties** panel consolidates several settings and editing features in a single location for easy access and use.

TIP The Properties panel appears by default in the Essentials Classic workspace.

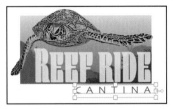

Properties panel controls

The available controls are dependent on the selected object and are organized by category:

- **Transform:** Dimensions, position, angle, etc.

- **Appearance:** Fill and stroke, opacity, effects, etc.

- **Dynamic:** Type settings, cropping, masking, etc.

- **Quick actions:** Tasks associated with the selection, such as creating outlines from text objects

Open the Properties panel

Do the following:

- Choose **Window** > **Properties Panel.**

Access full panel

Do the following:

- Click the **View More Options** button from the appropriate panel section to open the full panel (**Figure 2.14**).

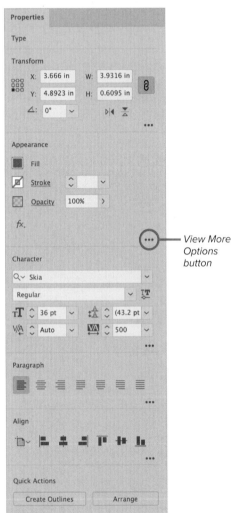

View More Options button

FIGURE 2.14 Displaying the controls for a selected text object in the Properties panel

Working with the Control Panel

TIP The Control panel appears by default in the Essentials Classic workspace.

The **Control** panel lets you quickly access settings for selected elements. By default, the **Control** panel resides docked at the top of the application frame (**Figure 2.15**).

Gripper bar   *Panel menu*

FIGURE 2.15 Displaying the attributes of a gradient object in the Control panel

Open the Control panel

Do the following:

- Choose **Window** > **Control Panel**.

Change the dock position

Do the following:

- From the **Control** panel menu, select **Dock to Bottom** or **Dock to Top**.

Float the Control panel

Do the following:

- Click+drag the gripper bar away from the docked location.

Customize which controls appear in the panel

Do the following:

- From the **Control** panel menu, select or deselect the settings you want to appear in the panel.

TIP The options displayed in the Control panel depend on the size of your application frame and the number of options selected in the panel menu.

Customizing Workspaces

Illustrator includes different project-based application configurations, as well as the capability to create and manage new workspaces.

Access the Workspace menu

Do either of the following:

- Choose **Window** > **Workspace**.
- Click the **Switch Workspace** button on the right side of the application bar (**Figure 2.16**).

Reset a workspace

If you change the configuration of a workspace (open or close a panel, add or delete a tool, etc.), you can revert to the original settings by doing either of the following:

- Choose **Window** > **Workspace** > **Reset** *[workspace name]*.
- Click the **Switch Workspace** button and select **Reset** *[workspace name]*.

Save a workspace

You can create a customized workspace using the application settings you've configured.

1. Customize the interface to suit your needs.
2. Choose **New Workspace** from the **Workspace** menu.
3. In the **New Workspace** dialog box, enter a name and then click **OK** (**Figure 2.17**).

FIGURE 2.16 The Switch Workspace menu

Customizing the workspace

These are some options for customizing the workspace:

- Open and close panels
- Dock and float panels
- Collapse and expand panels
- Add and delete tools in the toolbar
- Open or close the Control panel

FIGURE 2.17 Creating a saved workspace and the result displayed in the menu

Make a copy of a saved workspace

Do the following (**Figure 2.18**):

1. Choose **Manage Workspaces** from the **Workspace** menu.

2. In the dialog box, select the saved workspace you want to duplicate.

3. Click the **New Workspace** button.

4. Customize the name if you like and then click **OK**.

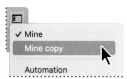

FIGURE 2.18 Creating a copy of a saved workspace and the result displayed in the menu

Rename a saved workspace

Do the following:

1. Choose **Manage Workspaces** from the **Workspace** menu.

2. In the dialog box, select the saved workspace you want to rename.

3. Edit the name.

4. Click **OK** to apply the change.

Delete a saved workspace

Do the following (**Figure 2.19**):

1. Choose **Manage Workspaces** from the **Workspace** menu.

2. In the dialog box, select the saved workspace you want to delete.

3. Click the **Delete Workspace** button.

4. Click **OK** to remove the workspace.

TIP You can copy, rename, or delete only workspaces that you create. The workspaces included with the application cannot be altered.

FIGURE 2.19 Deleting a saved workspace and the result displayed in the menu

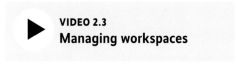

> ▶ **VIDEO 2.3**
> **Managing workspaces**

Configuring Illustrator Preferences

The **Preferences** panel lets you customize the settings for your Illustrator application (display options, commands, panel positions, type settings, etc.).

Open the Preferences dialog box

The **Preferences** dialog box allows you to customize your Illustrator application.

To access the dialog box (**Figure 2.20**), do any of the following:

- Choose **Edit** > **Preferences** (Windows) or **Illustrator** > **Preferences** (macOS) and select an option from the context menu.

- Click the **Preferences** button in the **Control** panel.

TIP To learn more about the individual Preferences tab sections, see Appendix A.

Set a preference

In the **Preferences** dialog box, do the following:

1. Select the appropriate section tab.

2. Modify the individual settings, as needed.

3. Click **OK** to apply the change.

Reset all preferences

To restore the default application preferences, do the following

1. In the **General** tab section of the **Preferences** dialog box, click the **Reset Preference** button.

2. Click **OK** to close the dialog box and confirm the reset.

3. Quit and then relaunch Illustrator to have the default preferences take effect.

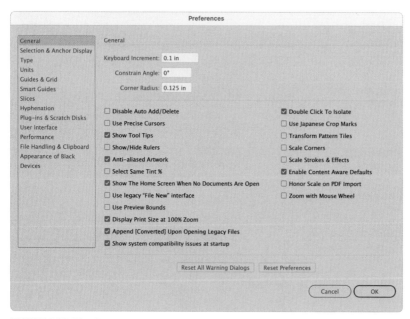

FIGURE 2.20 The General section of the Preferences dialog box

3

Working with Artboards

Artboards let you place designs on different size layouts, helping you visualize how they will appear on different size pages and devices.

Artboards are also useful for creating video storyboards or laying out animation elements.

In This Chapter

Artboards Overview

An artboard defines the area that contains your document's printable or exportable artwork (**Figure 3.1**).

Access an artboard's settings

Do either of the following:

- Select the **Artboard** tool from the toolbar.

- In the **Artboards** panel, click the **Artboard Options** icon.

TIP Once artboard features are activated, they are accessible in the Properties panel and Control panel as well at the Artboards panel.

Open the Artboards panel

The **Artboards** panel helps you create, manage, and navigate through artboards. To access the Artboards panel, do the following:

- Choose **Window** > **Artboards** (**Figure 3.2**).

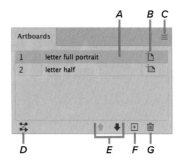

FIGURE 3.2
A. Active artboard **B.** Artboard Options icon
C. Panel menu **D.** Rearrange All Artboards button
E. Move Up / Move Down buttons
F. New Artboard button **G.** Delete Artboard button

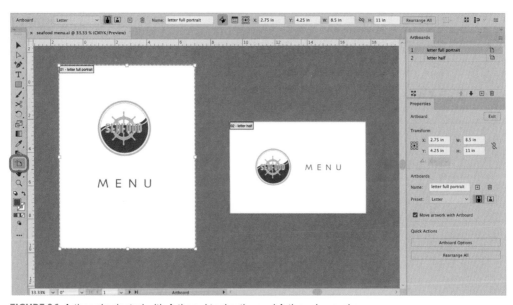

FIGURE 3.1 Artboard selected with Artboard tool active and Artboards panel open

TIP To learn about adding artboards to new documents, see "Creating a New File" in Chapter 1.

Adding Artboards

All Illustrator files must contain at least one artboard. You can create up to 1,000 artboards for each file.

Add using the Artboard tool

Do the following:

1. Select the **Artboard** tool in the toolbar (**Figure 3.3**).

2. In the document window, **click+drag** to define the size and position of the new artboard.

Add using the Properties panel or Control panel

Do the following:

- With artboard features active, click the **Add Artboard** icon (**Figure 3.4**).

FIGURE 3.3 Using the Artboard tool to manually create an artboard

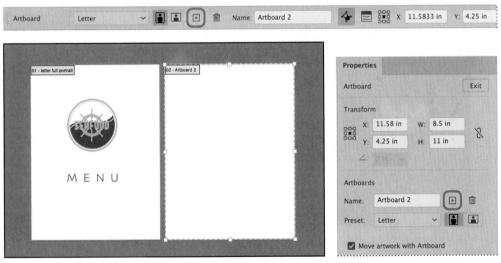

FIGURE 3.4 The Add Artwork icon in the Control and Properties panels

Add using the Artboards panel

In the Artboards panel, do either of the following:

- Click the **New Artboard** button (**Figure 3.5**).
- Click the panel menu and select **New Artboard**.

TIP The dimensions of artboards created using the Artboard panel are determined by the active artboard, or the top-level artboard if no artboard is active in the panel.

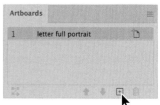

FIGURE 3.5 Clicking the New Artboard button in the Artboard panel to create an artboard

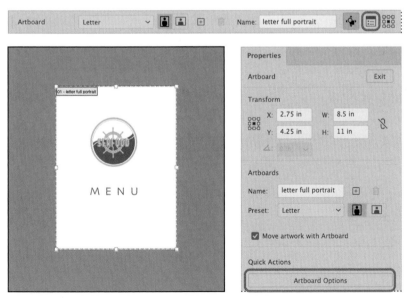

Modifying Artboards

Artboards can be modified in the **Art-boards Options** dialog box (**Figure 3.6**).

Open the Artboard Options dialog box for a selected artboard

Do any of the following:

- Double-click the **Artboard** tool.

- In the document window, double-click the artboard.

- In the **Artboards panel**, click the **Artboard Options** icon or select **Artboard Options** from the panel menu.

- From the **Control panel**, click the **Artboard Options** button (**Figure 3.7**).

- In the **Properties panel**, click **Artwork Options** under **Quick Actions**. (**Figure 3.7**).

FIGURE 3.6 The Artboard Options dialog box displaying the settings for the selected artboard

TIP An artboard must be selected to open its **Artboard Options dialog box.**

FIGURE 3.7 The Control panel and Properties panel with Artboard Options highlighted

Rename an artboard

Do either of the following:

- In the **Artboard Options** dialog box, enter a new name in the **Name** field, and then click **OK**.

- In the **Artboards** panel, double-click the artboard label and type a new name. Then, press **Enter** or **Return** to apply the change (**Figure 3.8**).

TIP Applying descriptive names to artboards is useful when working with multiple sizes.

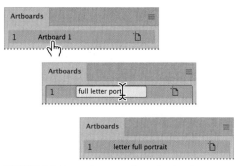

FIGURE 3.8 Renaming an artboard using the Artboards panel

Change dimensions using presets

Presets determine the dimensions of the artboard using the appropriate settings for the document type.

With the artboard selected, do any of the following:

- In the **Artboard Options** dialog box, select an option from the **Preset** menu (**Figure 3.9**).

- In the **Properties** panel, under the Artboards section, select an option from the **Preset** menu.

- From the **Control** panel, select an option from the **Preset** menu (**Figure 3.10**).

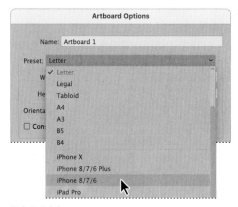

FIGURE 3.9 Selecting a preset in the Artboard Options dialog box

FIGURE 3.10 Preset menu in the Control and Properties panels

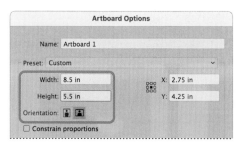

FIGURE 3.11 Dimension and orientation settings in the Artboard Options dialog box

Change dimension and orientation options

In either the **Artboard Option**s dialog box, **Properties** panel, or **Control** panel, do any of the following:

- Change the dimensions by entering new sizes in the **Width** (W) and **Height** (H) fields (**Figure 3.11**).

- Change the orientation by selecting either **portrait** or **landscape**.

Constrain dimension proportions

Do any of the following (**Figure 3.12**):

- In the **Artboard Options** dialog box, select **Constrain Proportions**.

- In the **Properties** panel or **Control** panel, click the **Maintain Width and Height Proportions** icon.

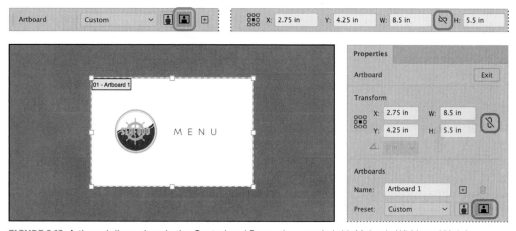

FIGURE 3.12 Artboard dimensions in the Control and Properties panels (with Maintain Width and Height Proportions deselected)

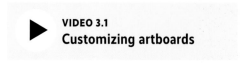

VIDEO 3.1
Customizing artboards

Change dimensions manually

With the artboard selected, do the following:

1. In the document window, select the artboard (**Figure 3.13**).

2. Hover over a corner or edge of the artboard boundary until the cursor displays as a **double arrow**.

3. **Click+drag** to adjust the dimensions.

FIGURE 3.13 Manually changing artboard dimensions in the document window

Reposition manually

With the artboard selected, do the following:

- In the document window, click inside the artboard boundaries and drag to a new position.

Set artboard display options

In the **Artboard Options** dialog box (**Figure 3.14**), do any of the following (**Figure 3.15**):

- Select **Show Center Mark** to display the center point.

- Select **Show Cross Hairs** to display lines crossing the center of each edge

- Select **Show Video Safe Areas** to display an artboard's boundary lines for the video's viewable area.

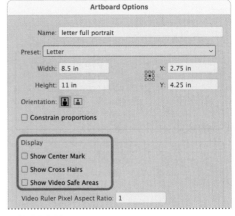

FIGURE 3.14 Artboard display options

FIGURE 3.15 Artboard display options

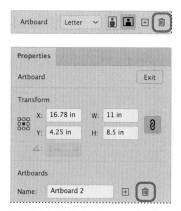

FIGURE 3.16 Delete icon in the Control and Properties panels

TIP Deleting an artboard does not remove its contents.

Managing Artboards

Organizing artboards helps maintain document efficiency.

Remove unused artboards

In the **Artboards** panel, do the following:

- Select **Delete Empty Artboards** from the panel menu.

Delete artboards

With the artboard selected, do any of the following:

- Press **Delete**.
- Click the **Delete Artboard** button in either the **Artboards** panel, **Properties** panel, or **Control** panel (**Figure 3.16**).
- In the **Artboards** panel, select **Delete** from the panel menu.

Duplicate an artboard and its contents

With the artboard selected, do the following:

1. Select **Move/Copy with Artwork** in either the **Control** or **Properties** panel.
2. In the **Artboards** panel, select **Duplicate Artboards** from the panel menu (**Figure 3.17**).

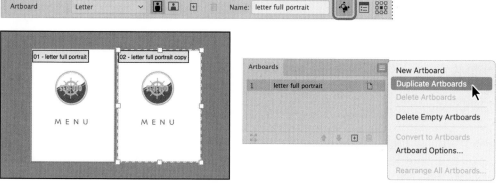

FIGURE 3.17 Move/Copy with Artwork selected in the Control panel while duplicating an artboard and its contents

Select multiple artboards

With an artboard selected, do any of the following (**Figure 3.18**):

- In the **Artboards** panel, **shift+click** to select additional artboards.

- In the document window, press **shift+click** to select additional artboards.

- In the document window, **click+drag** to select additional artboards.

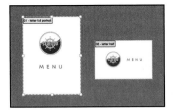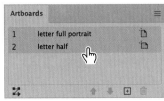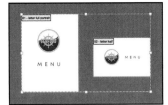

FIGURE 3.18 Selecting an additional artboard using the Artboards panel

Redistribute artboards

Do the following (**Figure 3.19**):

1. Either select **Rearrange All** in the **Properties** panel or click the **Rearrange All Artboards** button in the **Control** or **Artboards** panel.

2. In the dialog box, customize the artboard arrangement settings as needed.

3. Click **OK** to redistribute the artboards.

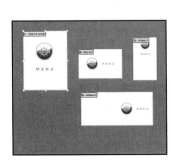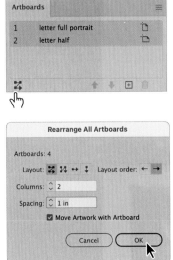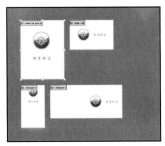

FIGURE 3.19 Rearranging a document's artboards using the Artboards panel

Working with Color

Illustrator provides a robust array of tools and controls for applying and managing colors to suit your project needs.

Accessing Fill and Stroke Controls

Fill is a color, pattern, or gradient inside a path or object.

Stroke is the visible outline of a path or object.

TIP **To learn more about working with strokes, see Chapter 8.**

Controls on the toolbar

The toolbar contains detailed fill and stroke control options (**Figure 4.1**).

Fill box shows the current fill.

Stroke box shows the current stroke.

Swap Fill and Stroke switches the fill and stroke colors.

Default Fill and Stroke changes the fill to white and the stroke to black.

Color is the last-selected solid color.

Gradient is the last-selected gradient.

None removes a fill or stroke from the selected object.

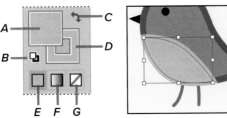

FIGURE 4.1
A. Fill box **B.** Default Fill and Stroke **C.** Swap Fill and Stroke **D.** Stroke box **E.** Color **F.** Gradient **G.** None

Select fill or stroke

Do the either of the following:

- To make fill the active choice, click the fill box.

- To make stroke the active choice, click the stroke box.

TIP **To learn more about gradients and patterns, see Chapter 13.**

Controls in other panels

Basic fill and stroke controls are also available in the **Control, Properties, Swatches,** and **Appearance** panels (**Figure 4.2**).

TIP **To access one of the panels with fill and stroke controls, choose Window > [panel name].**

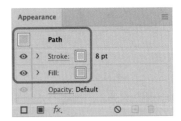

FIGURE 4.2 Fill and stroke controls in the Control, Properties, Swatches, and Appearance panels

---

Using the Eyedropper Tool

The **Eyedropper** tool lets you easily change the active or selected fills and strokes by sampling objects.

Update a document's active fill and stroke using the Eyedropper tool

1. Select the **Eyedropper** tool from the toolbar (**Figure 4.3**).

2. In the document window, click the object with the fill and stroke you want to be the active settings.

> **TIP** Objects do not need to be selected to be sampled using the Eyedropper tool.

FIGURE 4.3 Using the Eyedropper tool to update the active fill and stroke by sampling an object

> ▶ **VIDEO 4.1**
> **Applying fills and strokes using the Eyedropper tool**

Change an object's fill and stroke using the Eyedropper tool

1. In the document window, select the object with the fill and stroke you want change (**Figure 4.4**).

2. Select the **Eyedropper** tool.

3. In the document window, click the object with the fill and stroke you want to be the selected object's new settings.

A

B

C

FIGURE 4.4
A. Selecting the object to change **B.** Using the Eyedropper tool to sample the new fill and stroke
C. Updated fill and stroke in both the toolbar and the selection

Using the Color Picker

The **Color Picker** opens when you activate a change to a current color (**Figure 4.5**).

The **Color Picker** lets you accurately customize colors using four color mode options: **HSB**, **RGB**, **Hexadecimal**, and **CMYK**.

Correct color warnings

Out of Gamut Color means the color cannot be printed accurately (**F** in **Figure 4.5**). To correct this, do the following:

- Click the warning icon or the correction swatch to accept the substitute.

Out of Web Color means the color may not display correctly on all browsers or platforms (**G** in **Figure 4.5**). To correct this, do either of the following:

- Click either the warning icon or the correction swatch to accept the substitute.

- Select the **Only Web Colors** checkbox to reduce the color options to 256 web-safe color settings (**Figure 4.5**).

Color mode options

HSB (Hue, Saturation, and Brightness) is designed as a user-friendly approach to select colors.

- **Hue** is chosen in the color slider, which is based on the color wheel.

- **Saturation** and **Brightness** are specified as percentages.

RGB (Red, Green, and Blue) is designed for digital displays, making it the appropriate choice for any images that will be viewed onscreen.

(Hexadecimal) is an alternative notation of RGB colors. It's mainly used in web and screen design to describe colors in code, because the notation is more compact. (but it describes the same numbers).

CMYK (Cyan, Magenta, Yellow, and Black) was designed for traditionally printed files.

> **TIP** When using a color mode different than the document's, the color will be converted to the document color mode automatically. This is most striking when working in CMYK documents and specifying color as RGB.

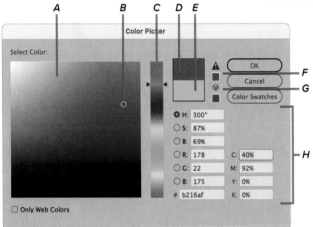

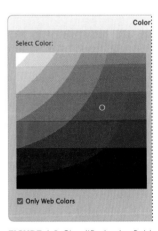

FIGURE 4.5
A. Color field **B.** Picked color **C.** Color slider **D.** Adjusted color
E. Original color **F.** Out of Gamut Color warning and correction
swatch **G.** Out of Web Color warning and correction swatch

FIGURE 4.6 Simplified color field with Only Web Colors selected

TIP Double-clicking the fill or stroke box opens the Color Picker even if a gradient or pattern is applied.

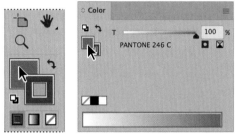

FIGURE 4.7 Double-clicking the fill box to open the Color Picker

TIP To learn more about swatches, see "Using the Swatches Panel" section in this chapter.

Access the Color Picker

Double-click the fill box or stroke box in any of the following (**Figure 4.7**):

- Toolbar
- **Swatches** panel
- **Color** panel

Apply a new active fill or stroke color using the Color Picker

Do any of the following, and then click **OK**:

- In the Color field, click to select the desired tint.
- Enter new color values in the appropriate fields.
- If Hue (**H**) is selected, use the **Color Slider** to adjust the color.
- Click **Color Swatches** to choose a document swatch as the new color (**Figure 4.8**).

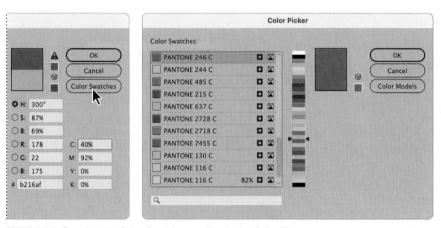

FIGURE 4.8 Opening the Color Swatches section in the Color Picker

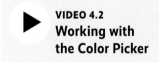

VIDEO 4.2
Working with the Color Picker

Using the Color Panel

The **Color** panel provides several controls and settings for modifying color type and mode settings, including:

- Fill and stroke settings
- Color mode value sliders and text boxes
- Color type conversion option

View color options

To view the **Color** panel's full controls and settings, do the following:

- Select **Show Options** from the panel menu (**Figure 4.9**).

Color types

Process colors are created by combining the four CMYK printing colors.

If you designate a process color's mode as RGB (or hexadecimal or HSB) in a CMYK document, it will be converted to CMYK immediately.

Spot colors are premixed inks used to replace CMYK mixtures.

The Pantone system is an industry standard for creating spot colors. Several Pantone spot color libraries are included with the Illustrator application.

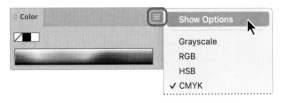
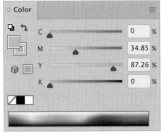

FIGURE 4.9 Showing the Color panel's options

Convert a spot color to a process color

Spot colors have limited editing capabilites. Converting them to process colors provides greater flexibility. Do the following:

- Click the **Spot Color** button (**Figure 4.10**).

TIP Spot colors should be used only when printing with the respective inks. When they are used in a file for this purpose, they should not be converted to process colors for convenience.

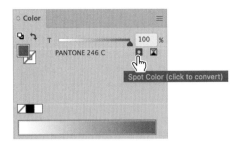
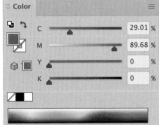

FIGURE 4.10 Converting a spot color to a CMYK process color

TIP Adjusting spot color value changes the tint, not the hue, and does not alter the opacity.

TIP Changing a color mode controls does not change the document color mode. To learn how to change the document color mode, see the "Apply Color Changes Using Menus" section in this chapter.

Adjust a color value

Do either of the following:

- Click+drag the color value slider.
- Enter new color values in the text boxes.

Change the color mode controls

Do the following (**Figure 4.11**):

1. In the **Color** panel, click the panel menu button.

2. Select a different color mode view.

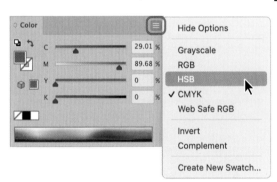
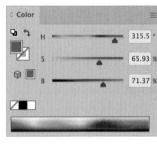

FIGURE 4.11 Changing the Color panel's color mode view from CMYK to HSB

TIP Spot colors must be converted to process colors before applying complement or inverse actions.

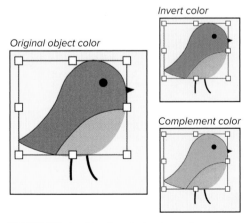

FIGURE 4.12 Applying inverted and complement colors to an object

Change an object's process color to its complement or inverse

With the object selected (**Figure 4.12**), do the following:

1. Select the fill or stroke process color box you want to change.

2. In the **Color** panel, click the panel menu and select either of the following:

   Invert changes a color to its opposite value.

   Complement changes a color to a value calculated using its lowest and highest RGB values.

Using the Swatches Panel

The **Swatches** panel (**Figure 4.13**) is an essential tool for organizing a document's colors, tints, patterns, and gradients.

TIP To learn more about gradients and patterns, see Chapter 13.

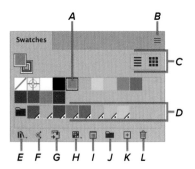

FIGURE 4.13
A. Active swatch **B.** Panel menu
C. Show List / Thumbnail View buttons
D. Color group **E.** Swatch libraries
F. Open Color Themes panel
G. Add Swatch to My Library
H. Show Swatch Kinds menu
I. Swatch Options **J.** New Color Group
K. New Swatch **L.** Delete Swatch

View swatch details

To change from thumbnail to list view to see the names of the swatches, do either of the following:

- Click the **Show List View** button.

- From the panel menu (**Figure 4.14**), select a list view option.

Change swatch display size

To increase or decrease the size of the swatches in either list or thumbnail view, do the following:

- Select a swatch size view option from the panel menu.

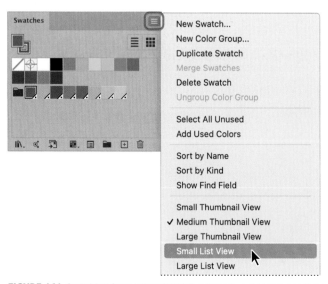

FIGURE 4.14 Switching from a thumbnail to a list view using the panel menu

Swatch color types

The **Swatches** panel may contain different combinations of swatch color types and uses different thumbnails to identify them.

 The **Registration** color is a built-in, nonremovable swatch used for printer marks. Objects that are assigned the registration swatch print on every separation when output to a PostScript printer.

 Process colors are a combination of CMYK (the four standard printing inks: cyan, magenta, yellow, and black). By default, Illustrator defines new swatches (including RGB) as process colors.

 Global Process colors automatically update throughout your artwork when you modify them. Global color swatches are identifiable by the triangle in the lower corner of the icon.

 Spot colors are premixed inks used to replace CMYK mixtures. Spot-color swatches are identifiable by the dot inside the triangle in the lower corner of the icon.

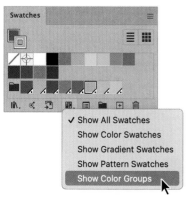

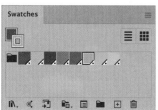

FIGURE 4.15 Showing color group swatches (and hiding all others) using the Show Swatch Kinds menu

Display certain swatch types

In the **Swatches** panel, do the following:

- Click the **Show Swatch Kind** button and select an option (**Figure 4.15**).

Arrange swatches by type

In the **Swatches** panel, do the following:

- From the panel menu, select a **Sort By** option to arrange the swatches by name or type.

Create a swatch from an active fill or stroke color

In the **Swatches** panel, do the following:

- Click+drag the color box onto the swatches section (**Figure 4.16**).

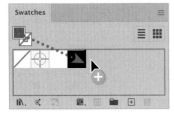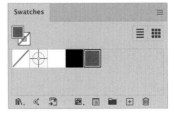

FIGURE 4.16 Dragging an active color to create a new swatch

Create a swatch using customized settings

In the **Swatches** panel, do the following (**Figure 4.17**):

1. Either click the **New Swatch** button or select **New Swatch** from the panel menu.

2. In the **New Swatch** dialog box, customize the settings by doing any of the following, and then click **OK**:

   Change the **Swatch Name**.

   Change the swatch **Color Type**.

   Adjust the color settings using the **Color Mode** options.

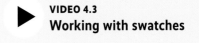

VIDEO 4.3
Working with swatches

TIP Hovering over a swatch thumbnail displays the name.

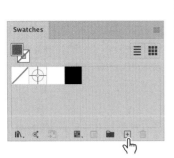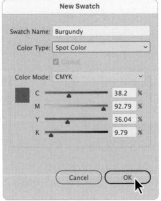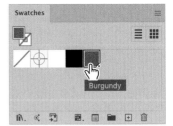

FIGURE 4.17 Creating a customized global spot color swatch from a process color

Create a swatch using the Color panel

In the **Color** panel, do the following (Figure 4.18):

1. Adjust the active color, as needed.

2. Select **Create New Swatch** from the panel menu.

3. In the **New Swatch** dialog box, customize the settings as needed and then click **OK**.

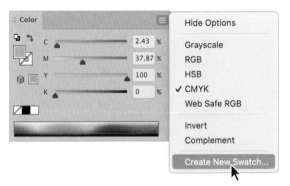
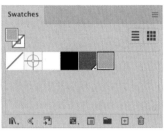

FIGURE 4.18 Creating a swatch using the Color panel menu

Modify a swatch

In the **Swatches** panel, do the following (Figure 4.19):

1. Either **double-click** the swatch thumbnail or choose **Swatch Options** from the panel menu with the active swatch selected.

2. In the **Swatch Options** dialog box, customize the settings by doing any of the following and then click **OK**:

Change the swatch **Name**.

Change the swatch **Color Type**.

Adjust the color settings using the **Color Mode** options.

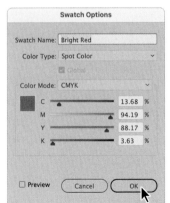
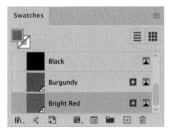

FIGURE 4.19 Modifying swatch options by double-clicking the thumbnail

Create a color group
from selected swatches

Do the following:

1. Either click the **New Color Group** button (**Figure 4.20**) or select **New Color Group** from the panel menu.

2. In the **New Color Group** dialog box, enter a **Name** for the group and click **OK**.

TIP **To add swatches to a color group, drag them onto the group.**

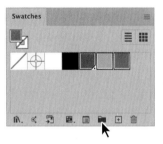 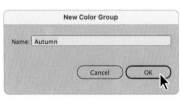 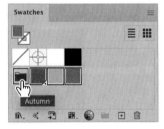

FIGURE 4.20 Creating a new color group from selected swatches and the result

Create a color group
from selected objects

Do the following:

1. Either click the **New Color Group** button (**Figure 4.21**) or select **New Color Group** from the panel menu.

2. In the **New Color Group** dialog box, enter a **Name** for the group and then click **OK**.

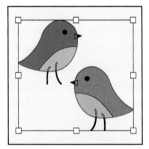

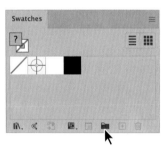  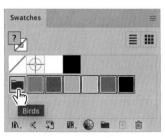

FIGURE 4.21 Creating a new color group from selected objects and the result

Access swatch libraries

Illustrator provides a robust collection of ink-based and thematic color libraries.

To access a swatch library, do either of the following:

- Click the **Swatch Library** button and select an option from the menu.
- Select **Open Swatch Library** > *[library name]* from the panel menu (**Figure 4.22**).

Add library swatches to the Swatches panel

With the library swatch(es) selected, do either of the following:

- Click a library swatch.
- Drag the swatch(es) from the library panel onto the **Swatches** panel.
- Select **Add to Swatches** from the library panel menu.

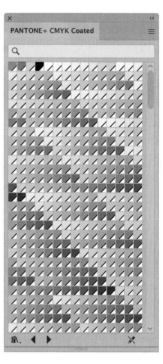

FIGURE 4.22 Opening a swatch library using the Swatch library button

Create a custom swatch library

With the **Swatches** panel organized as needed, do the following:

1. Click the **Swatch Library** button and select **Save Swatches**.

2. In the **Save Swatches as Library** dialog box, enter a **Name** and select a location for the library; then click **Save**.

TIP The default location for swatch libraries is **Illustrator/Presets/Swatches**.

Using the Color Guide Panel

The **Color Guide** panel (**Figure 4.23**) provides harmonious color variation suggestions based on the current color.

TIP To learn more about the Edit Colors button, see the "Apply Color Changes Using Menus" section in this chapter.

A   *B*

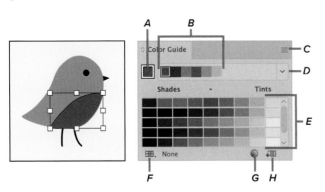

FIGURE 4.23
A. Set as Base Color
B. Active color group
C. Panel menu
D. Harmony Rules menu
E. Color variations
F. Limits colors to specified swatch library
G. Edit Colors
H. Save Group to Swatches panel

Set the Color Guide base color

Do either of the following:

- Use the **Eyedropper** tool to sample a color in the document

- Select a swatch from the **Swatches** panel.

Change the Color Guide options

Do the following:

1. Select **Color Guide Options** from the panel menu (**Figure 4.24**).

2. Adjust the **Steps** and **Variation** as needed; then click **OK**.

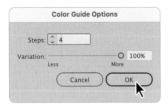

FIGURE 4.24 The Color Guide Options dialog box

Apply a color variation to a selected object's fill

Do either of the following:

- Click an **Active Color Group** swatch.

- Click a **Color Variations** swatch.

Change the harmony rule

Do the following:

- Click the **Harmony Rules** menu button (**Figure 4.25**) and choose a new rule.

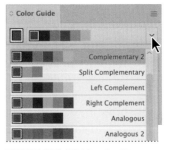

FIGURE 4.25 Clicking the Harmony Rule menu to view the options

Change the color options using swatch libraries

- Click the **Limit the Colors to a Specified Swatch Library** button and select an option from the menu (**Figure 4.26**).

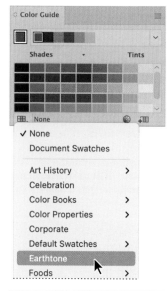

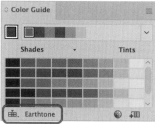

FIGURE 4.26 Selecting a swatch library to limit the Color Guide suggestions

Change the type of color variation

Select any of the following from the panel menu:

- **Show Tints/Shades** adds black (tint) and white (shades) to the variations.
- **Show Warm/Cool** adds red (warm) and blue (cool) to the variations.
- **Show Vivid/Muted** increases (vivid) and decreases (muted) saturation.

Add a color group to the Swatches panel

Do the either of the following:

- Click the **Add Group to Swatches Panel** button (**Figure 4.27**).
- Select **Save Colors as Swatches** from the panel menu.

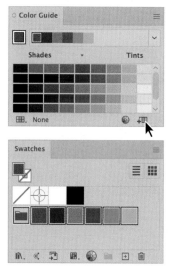

FIGURE 4.27 Adding the active color group to the Swatches panel

Applying Color Changes Using Menus

Change the color mode for an entire document

With the document open, select the appropriate color mode for your project by doing either of the following (**Figure 4.28**):

- Choose **File** > **Document Color Mode** > **RGB Color** for digital projects.
- Choose **File** > **Document Color Mode** > **CMYK Color** for printed projects.

FIGURE 4.28 Selecting a new color mode for a document using the File menu

Change the color mode for objects in CMYK documents

With the objects selected, do either of the following:

- Choose **Edit** > **Edit Colors** > **Convert to RGB** to convert CMYK objects (**Figure 4.29**).
- Choose **Edit** > **Edit Colors** > **Convert to CMYK** to convert RGB objects.

TIP When the document color mode is changed, the embedded RGB objects are converted to CMYK automatically and vice versa.

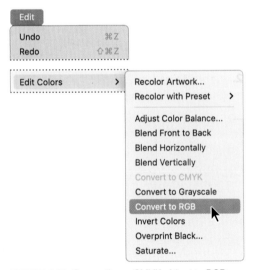

FIGURE 4.29 Converting a CMYK object to RGB

Change objects to their inverse color

With the objects selected, do the following (Figure 4.30):

1. Make sure all selected objects are assigned process colors.

2. Choose **Edit** > **Edit Colors** > **Invert Colors**.

Change object colors to grayscale

With the objects selected, do the following (**Figure 4.31**):

1. Make sure all selected objects are assigned process colors.

2. Choose **Edit** > **Edit Colors** > **Convert to Grayscale**.

Change the color saturation of objects

Do the following:

1. Select the objects you want to change and choose **Edit** > **Edit Colors** > **Saturate**.

2. In the dialog box, adjust the **intensity** and then click **OK** (**Figure 4.32**).

FIGURE 4.32 Desaturating selected objects

Change the color balance of objects

Do the following:

1. Select the objects you want to change and choose **Edit** > **Edit Colors** > **Adjust Color Balance**.

2. In the dialog box, adjust the color settings for the fill and/or stroke, and then click **OK** (**Figure 4.33**).

FIGURE 4.30 Inverting colors of selected objects

FIGURE 4.31 Converting the colors for selected objects to grayscale

TIP Spot and global colors cannot be inverted or changed to grayscale.

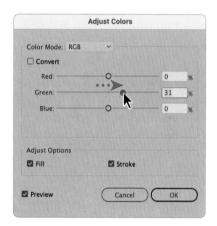

FIGURE 4.33 Adjusting the color balance for selected objects by increasing the green RGB value

Edit or recolor artwork

To open the **Edit Colors / Recolor Artwork** dialog box, do the following (**Figure 4.34**):

1. Select the objects you want to edit or recolor.

2. Choose **Edit** > **Edit Colors** > **Recolor Artwork**.

3. Adjust the color settings as needed using the options in the dialog box.

TIP The dialog box name and appearance varies depending on how it is activated. These controls are also available in the Control panel, Color Guide panel, and Swatches panel (if a color group is selected).

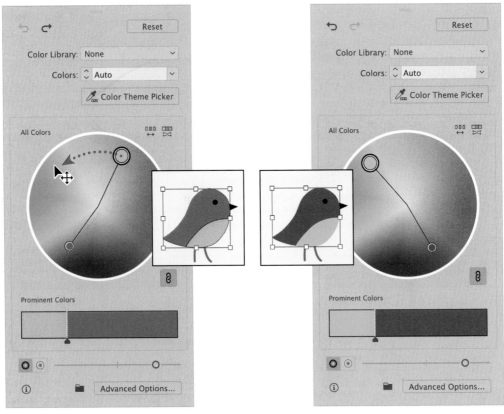

FIGURE 4.34 Recoloring selected objects

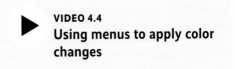

VIDEO 4.4
Using menus to apply color changes

Navigating Documents

The Adobe Illustrator navigational tools and features let you focus on specific areas and elements in your document, helping you work efficiently.

Changing Magnification

Illustrator lets you easily change the document magnification levels using a variety of options.

Depending on your system's graphics processing unit (GPU), performance enhancements let Illustrator pan, zoom, and scroll up to 10 times faster with 10 times higher magnification (64,000%, up from 6,400%).

TIP To learn more about system requirements for GPU performance, click the **More Info** link in the Preferences dialog box, which will take you to the appropriate information in the online Adobe Illustrator User Guide.

Enable GPU performance

Do the following:

1. Open the **Preferences** dialog box by choosing either:

   Illustrator > **Preferences** > **Performance** (macOS)

   or

   Edit > **Preferences** > **Performance** (Windows)

2. Select the **GPU Performance** checkbox (**Figure 5.1**).

3. Click **OK**.

4. Choose **View** > **View Using GPU**.

Toggle between GPU and CPU view modes

Do any of the following:

- Press **Command/Ctrl+E**.
- Choose **View** > **View Using CPU**.
- Choose **View** > **View Using GPU**.

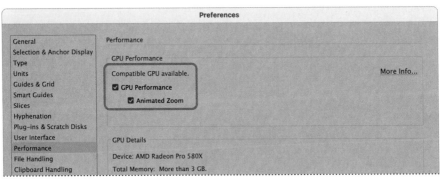

FIGURE 5.1 GPU Performance settings in the Preferences dialog box

FIGURE 5.2 Zoom tool in the toolbar

FIGURE 5.3 Using animated zoom to increase magnification by click+dragging the Zoom tool to the right

FIGURE 5.4 Using animated zoom to decrease magnification by click+dragging the Zoom tool to the left

TIP When animated zoom is disabled, the Zoom tool uses marquee zoom features instead.

FIGURE 5.5 Using marquee zoom to increase magnification by click+dragging the Zoom tool

Toggle between animated and marquee zoom

Do any of the following:

- Toggle the **Animated Zoom** checkbox in the **Performance** section of the **Preferences** dialog box.

- Choose **View** > **View Using CPU** to disable animated zoom.

- Choose **View** > **View Using GPU** to enable animated zoom.

Change magnification using animated zoom

With the **Zoom** tool active (**Figure 5.2**), do any of the following:

- To zoom in, either:

  Click+hold the cursor.

  Click+drag to the right (**Figure 5.3**).

- To zoom out, either:

  Click+hold the cursor while pressing Alt/Option.

  Click+drag to the left (**Figure 5.4**).

Increase magnification using marquee zoom

With the **Zoom** tool selected in the toolbar, do the following:

- Click+drag over the area (**Figure 5.5**).

Change magnification incrementally using the Zoom tool

Do either of the following:

- Zoom in by clicking the area you want to magnify.

- Zoom out by pressing **Alt/Option** while clicking the area you want to expand.

Change magnification using the View menu

Do any of the following (**Figure 5.6**):

- Choose **View** > **Zoom In** to incrementally increase magnification.

- Choose **View** > **Zoom Out** to incrementally decrease magnification.

- Choose **View** > **Fit Artboard in Window** to increase or decrease the magnification so the active artboard fits in the document window.

- Choose **View** > **Fit All in Window** to increase or decrease the magnification so all the document artboards fit in the document window.

- Choose **View** > **Actual Size** to display a print preview of your document at 100%.

Change magnification using the status bar

In the status bar located in the lower left of the document window, do either of the following (**Figure 5.7**):

- Enter a new percentage in the magnification field.

- Click the magnification menu button and select a percentage.

TIP Keyboard shortcuts for menu items are displayed to the right of the command.

FIGURE 5.6 Magnification options in the View menu

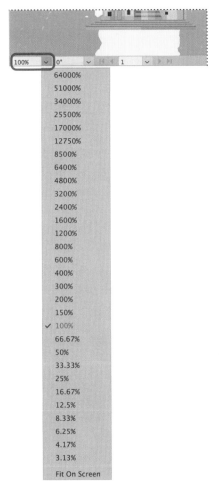

FIGURE 5.7 Magnification options in the status bar

TIP Pressing the F key toggles through the screen modes.

TIP Pressing the Esc key exits full screen or presentation modes and returns to Normal Screen mode.

FIGURE 5.8 Selecting a new screen mode using the toolbar

TIP Screen mode options are also available in the View menu.

Change screen modes using the toolbar

At the bottom of the toolbar, click the **Change Screen Mode** button and select any of the following (**Figure 5.8**):

- **Presentation Mode** displays the artwork while hiding the application menu, panels, and guides.

- **Normal Screen Mode** is the default working mode, with a menu bar at the top and scroll bars on the sides.

- **Full Screen Mode With Menu Bar** fills the screen and keeps the menu bar and scroll bars visible.

- **Full Screen Mode** displays artwork in a full-screen window, with no title bar or menu bar.

Using the Hand and Rotate View Tools

Illustrator lets you easily change the viewing area using a variety of options.

Pan to another area using the Hand Tool

With the **Hand** tool selected (**Figure 5.9**), do the following:

- Click+drag to move to another portion of the document window (**Figure 5.10**).

Change an artboard's orientation using the Rotate View Tool

With the **Rotate View** tool selected (**Figure 5.9**), do the following:

- Click+drag to change the orientation of the artboard (**Figure 5.11**).

Restore an artboard to the original orientation

Do the following:

- Double-click the **Rotation View** tool.

Pan to another area using the document scroll bar

Do either of the following:

- Click+drag the scroll button (**Figure 5.12**).
- Click either side of the scroll button to incrementally pan.

TIP The Hand tool can be toggled on when another tool is active by pressing and holding the spacebar.

FIGURE 5.9 The Hand tool group in the toolbar

TIP To learn about the Print Tiling tool, see Chapter 17.

FIGURE 5.10 Panning a document view using the Hand tool

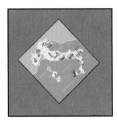

FIGURE 5.11 Changing the orientation of a document's view using the Rotate View tool

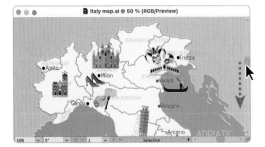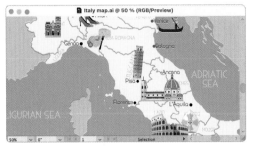

FIGURE 5.12 Panning down a document using the scroll button

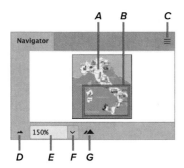

FIGURE 5.13
A. Artwork thumbnail **B.** Proxy view area
C. Panel menu **D.** Zoom Out button
E. Zoom box **F.** Zoom box menu
G. Zoom In button

Change magnification using the Navigator panel

Do any of the following:

- Click the **Zoom In** button to increase magnification (**Figure 5.14**).

- Click the **Zoom Out** button to decrease magnification.

- Enter a new magnification percentage in the **Zoom** box.

- Click the **Zoom** box menu and select a new magnification percentage.

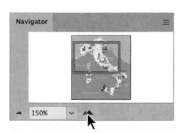

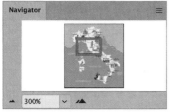

FIGURE 5.14 Increasing magnification using the Zoom In button

Using the Navigator Panel

The **Navigator** panel (**Figure 5.13**) uses a thumbnail display to let you quickly change the position and magnification of your document view.

The current document view is represented by an outlined color box called the *proxy view area*.

Pan to another area using the Navigator panel

Do the following:

- Click+drag to move to another portion of the artwork thumbnail (**Figure 5.15**).

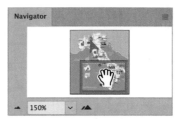

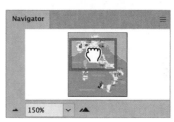

FIGURE 5.15 Panning a document view using the Navigator panel

Changing View Modes

By default, Illustrator displays artwork in Preview (full-color) mode. However, sometimes it's helpful to view only the paths of objects in Outline mode or use Isolation mode for easier selection.

Toggle between Preview and Outline views

Do any of the following:

- Choose **View** > **Preview** to display the art in full color (**Figure 5.16**).

- Choose **View** > **Outline** to display only the object paths (**Figure 5.17**).

TIP In Outline view, text elements display with black fill rather than paths.

- Press **Command/Ctrl+Y** to toggle between Preview and Outline modes.

Activate Isolation mode for an object or group

Isolation mode helps with editing selected objects, paths, and groups. The isolated elements appear in full color and the other elements are dimmed and not selectable.

To activate Isolation mode, do any of the following:

- With the **Selection** tool, double-click an object or group (**Figure 5.18**).

- With an object or group selected, click **Isolate Selected Object** in the **Control** panel (**Figure 5.19**).

- In the **Layers** panel, select the objects or groups and choose **Enter Isolation Mode** from the panel menu.

TIP To learn more about the Layers panel, see Chapter 6.

FIGURE 5.16 Preview mode

FIGURE 5.17 Outline mode

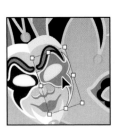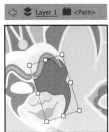

FIGURE 5.18 Double-clicking an object to activate Isolation mode (with the Isolation mode bar appearing above the document window)

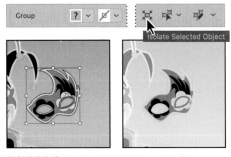

FIGURE 5.19 Activating Isolation mode for a selected group by clicking the Isolate Selected Object button in the Control panel

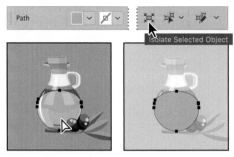

FIGURE 5.20 Activating Isolation mode for a path within a group using the Direct Selection tool

TIP To learn more about the Direct Selection tool, see Chapter 7.

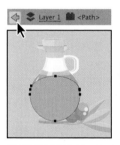

FIGURE 5.21 Exiting Isolation mode for a sublayer object using the Exit Isolation Mode button

Activate Isolation mode for paths within a group

Do the following (**Figure 5.20**):

1. Select the path using the **Direct Selection** tool or the **Layers** panel.

2. Click the **Isolate Selected Object** button in the **Control** panel.

Exit Isolation mode

Do any of the following:

- Press **Esc**.

- Click anywhere along the **Isolation Mode** bar.

- Double-click outside of the isolated group with the **Selection** tool.

- Click the **Exit Isolation Mode** button on the **Isolation Mode** bar one or more times (**Figure 5.21**).

TIP If a sublayer has been isolated, the first click takes you back a level. Depending on the number of sublevels, it may take the multiple clicks to exit Isolation mode.

VIDEO 5.1
Using view options to assist with selecting elements

Using Rulers

Rulers help you precisely position elements. They display along the top and left edges of the document window (**Figure 5.22**).

Show or hide rulers

Do any of the following:

- Choose **View** > **Rulers** > **Show Rulers**.
- Choose **View** > **Rulers** > **Hide Rulers**.
- Press **Command/Ctrl+R** to toggle between showing and hiding rulers.

Change a document's unit of measurement

Do either of the following:

- Right-click the ruler and choose a new unit for measure from the menu (**Figure 5.23**).
- Choose **File** > **Document Setup** and select a new option from the **Units** menu in the **General** section.

FIGURE 5.22 Rulers displayed in the top and left corners of the document window

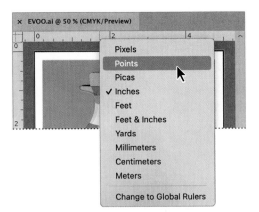

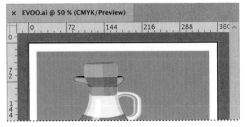

FIGURE 5.23 Changing the unit of measure by right-clicking the ruler

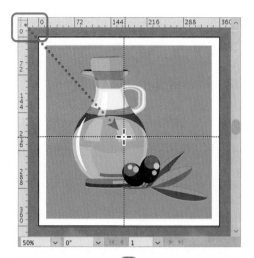

FIGURE 5.24 Changing the point of origin by click+dragging from the ruler intersection point

Change the ruler point of origin

The point of origin is the beginning position (0) for each ruler. To change it, do the following:

- Click+drag from the ruler intersection point to the new location (**Figure 5.24**).

Reset the ruler point of origin

Do the following:

- Double-click the ruler intersection point in the upper-left corner of the document window.

Artboard and Global rulers

Illustrator provides two different types of rulers:

- **Artboard** rulers change the point of origin based on each artboard. Artboard rulers are the default in Illustrator.

- **Global** rulers use a single point of origin for the entire document.

Toggle between Artboard and Global rulers

Do either of the following:

- Choose **View** > **Rulers** > **Change to Global Rulers**.

- Choose **View** > **Rulers** > **Change to Artboard Rulers**.

Using Guides and Grids

Guides and grids are nonprinting elements that appear on top of the artboard and can assist with precisely positioning objects.

Add a linear guide

With the rulers visible in the document window, do the following:

- Click+drag away from the horizontal or vertical ruler to the desired position (Figure 5.25).

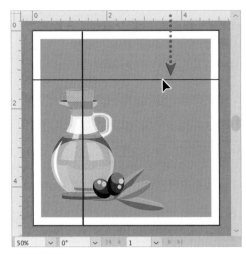

FIGURE 5.25 Creating linear guides by click+dragging away from the rulers

Toggle locking or unlocking guides

Do either of the following:

- Choose **View** > **Guides** > **Lock Guides**.
- Choose **View** > **Guides** > **Unlock Guides**.

Toggle guide visibility

Do the either of the following:

- Choose **View** > **Guides** > **Show Guides**.
- Choose **View** > **Guides** > **Hide Guides**.

Create guides from objects

Do the following (**Figure 5.26**):

1. Select the object to convert to a guide.
2. Choose **View** > **Guides** > **Make Guides**.

FIGURE 5.26 Creating a guide from a rectangle object

Convert guides back to objects

Do the following:

1. Make sure guides are unlocked.
2. Choose **View** > **Guides** > **Release Guides**.

Toggle guide snapping

To snap objects to guides, do the following:

- Choose **View** > **Snap To Point**.

Delete a guide

Do the following:

1. Make sure guides are unlocked.
2. With the **Selection** tool, click the guide and press **Delete** or **Backspace**.

Delete all guides

Do the following:

- Choose **View** > **Guides** > **Clear Guides**.

FIGURE 5.27 Grid visible in the document window

Toggle grid visibility

Do the either of the following:

- Choose **View** > **Show Grid** (**Figure 5.27**).
- Choose **View** > **Hide Grid**.

Toggle grid snapping

To snap objects to the grid, do the following:

- Choose **View** > **Snap to Grid**.

Set Preferences for guides and grids

Do the following (**Figure 5.28**):

- In the **Guides & Grid** section of the **Preferences** dialog box, adjust the settings as needed and then click **OK**.

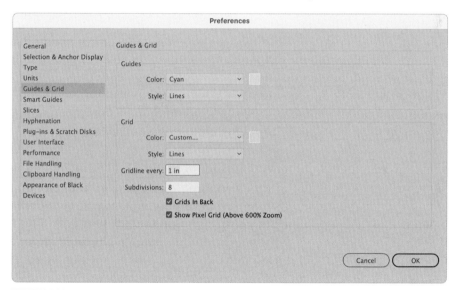

FIGURE 5.28 The Guides & Grid section of the Preferences dialog box

Using Smart Guides

Smart Guides appear automatically in Illustrator as you move or resize a selected element. They help you align, edit, and transform elements by snapping (aligning) to other elements, guides, and grids and by displaying the element's coordinates (**Figure 5.29**).

Set Preferences for Smart Guides

Do the following (**Figure 5.30**):

- In the **Smart Guides** section of the **Preferences** dialog box, adjust the display options and snapping tolerance as needed and then click **OK**.

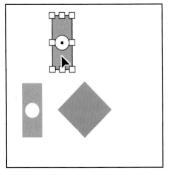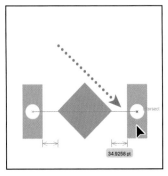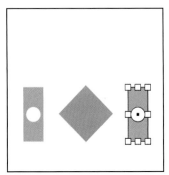

FIGURE 5.29 Repositioning an object using Smart Guides

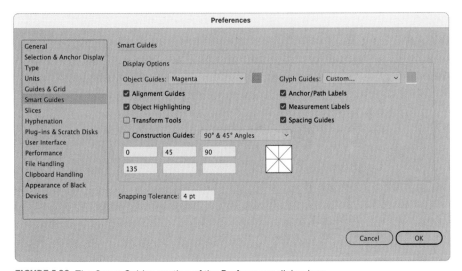

FIGURE 5.30 The Smart Guides section of the Preferences dialog box

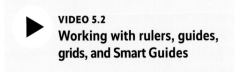

VIDEO 5.2
Working with rulers, guides, grids, and Smart Guides

6

Organizing Artwork

The Illustrator Layers panel and other tools help you manage all the elements that make up your artwork, enabling you to work efficiently and avoid frustration.

In This Chapter

Organizing Elements Using Layers

Layers are similar to stacked transparent folders and sheets that contain your document's artwork (**Figure 6.1**).

Using the **Layers** panel (**Figure 6.2**), you can organize objects, text, and other elements for easy access. How simple or complex a document's layer organizational structure should be is up to you—and the complexity of your artwork.

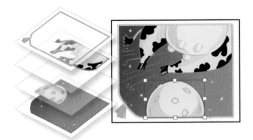

FIGURE 6.1 Layers comprising a document's artwork

Set layer options

For a layer or sublayer, do any of the following (**Figure 6.3**):

- Click the **Visibility** column (**A** in Figure 6.3) to show or hide it.

- Click the **Edit** column (**B** in Figure 6.3) to lock or unlock it.

- Click the **Target** column (**C** in Figure 6.3) to mark
it for editing or applying effects.

- Click the **Selection** (**D** in Figure 6.3) column to select the object within it.

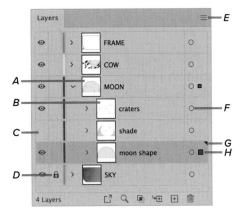

FIGURE 6.2
A. Expanded layer **B.** Sublayer
C. Hidden sublayer **D.** Locked layer
E. Panel menu **F.** Target icon
G. Active Layer for Next Action icon
H. Selected Object icon

Expand a layer to see its contents

In the **Layers** panel, do the following:

- Click the > icon to the left of the layer thumbnail.

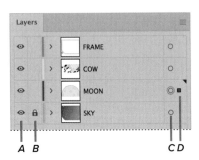

FIGURE 6.3
A. Visibility column **B.** Edit column
C. Target column **D.** Selection column

Organizing Complex Elements

By default, all of a document's elements reside as individual elements within a single parent layer. For complex documents, organizing the elements in a cohesive manner helps you work more efficiently and avoid frustration (**Figure 6.4**).

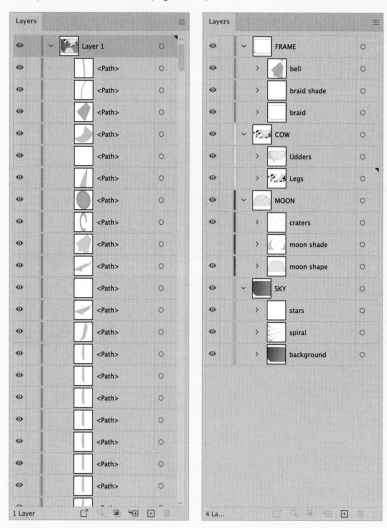

TIP Layers panel display options, including the size of the thumbnails, can be customized under Panel Options in the panel menu.

FIGURE 6.4 Layers panel before and after organizing a complex document

Add a layer using default settings

In the **Layers** panel, do the following (**Figure 6.5**):

1. Select the layer that you want the new layer to reside above.
2. Click the **Create New Layer** button.

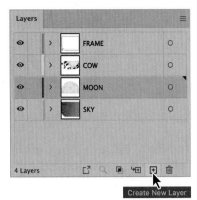

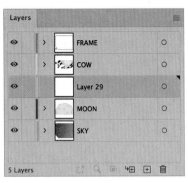

FIGURE 6.5 Adding a new layer using the Create New Layer button

TIP New elements will be automatically added within the active layer.

Add a layer using customized settings

In the **Layers** panel, do the following (**Figure 6.6**):

1. Select the layer that you want the new layer to reside above.
2. Choose **New Layer** from the panel menu.
3. In the **Layer Options** dialog box, customize the new layer settings (such as **Name** and **Color**) as needed, and then click **OK**.

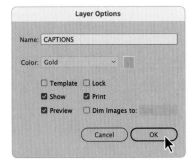

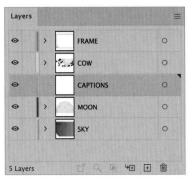

FIGURE 6.6 Adding a new customized layer using the Layer Options dialog box

Add a sublayer using default settings

In the **Layers** panel, do the following: (**Figure 6.7**):

1. Select the layer that you want the new sublayer to reside inside.
2. Click the **Create New Sublayer** button.

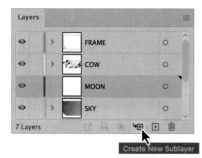

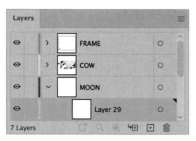

FIGURE 6.7 Adding a new sublayer using the Create New Sublayer button

Modify a layer or sublayer's properties

In the **Layers** panel, do the following:

1. Double-click the layer or sublayer you want to modify.
2. In the **Layer Options** dialog box, customize the settings (such as name and color) as needed and then click **OK**.

Add a sublayer using customized settings

In the **Layers** panel, do the following (**Figure 6.8**):

1. Select the layer that you want the new sublayer to reside inside.
2. Choose **New Sublayer** from the panel menu.
3. In the **Layer Options** dialog box, customize the new layer settings (such as **Name** and **Color**) as needed and then click **OK**.

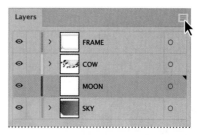

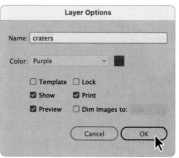

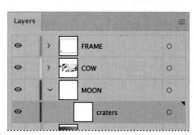

FIGURE 6.8 Adding a new customized sublayer using the Layer Options dialog box

Selecting Elements Using Layers

The **Layers** panel's hierarchal structure is a useful tool for selecting elements

When you select a layer or sublayer, all the elements beneath it are included as selections.

Select all the elements within a layer or sublayer

In the **Layers** panel, do the following:

1. Click the layer's selection area, located to the right of the target icon (**Figure 6.9**).

2. (Optional) Expand the layer or sublayer to see the selected content.

TIP In Illustrator, elements can be paths, groups, raster images, meshes, and so on.

Select an element within a layer or sublayer

In the **Layers** panel, do the following:

-   Click the element's selection area, located to the right of the target icon.

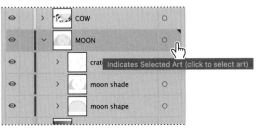

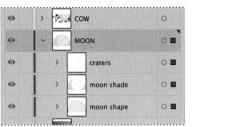

FIGURE 6.9 Clicking the selection area to select all the layer's elements within it

Select elements within different layers or sublayers

In the **Layers** panel, do the following:

1. Click the selection area for the first element.

2. Shift+click the selection area for additional elements to add them as selections.

Deselect elements

In the **Layers** panel, do the following:

-   Shift+click the selection square of the layer, sublayer, or element you want to deselect.

TIP You can select multiple consecutive elements on a layer by clicking the target icon of the first one and shift+clicking the target icon of the last one.

Group elements

Do the following (**Figure 6.10**):

1. Select the elements using the **Layers** panel or **Selection** tool.

2. Choose **Object** > **Group**.

> **TIP** Grouping elements from different layers places them in the topmost selected element's layer.

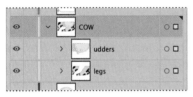

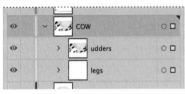

FIGURE 6.10 Grouping selected sublayer elements and the result

> **TIP** Deleting or cutting elements does not delete their layer. To learn about removing layers, see "Managing Artwork Hierarchy and Structure" in this chapter.

Ungroup elements

Do the following:

1. Select the group using the **Layers** panel or **Selection** tool.

2. Choose **Object** > **Ungroup**.

Working with Groups

Grouped elements are successively stacked in the same layer and treated as a single entity. Any modifications to the selected group (moving, scaling, color changes, and so on) are applied to all the elements.

Add elements to an existing group

With the group selected, do the following (**Figure 6.11**):

1. Enter isolation mode by either double-clicking the group with the **Selection** tool or choosing **Enter Isolation Mode** from the **Layers** panel menu.

> **TIP** For more information about selecting elements in Isolation mode, see Chapter 7.

2. Add elements by either pasting or drawing them.

FIGURE 6.11 Adding elements to a group by pasting them in Isolation mode

Managing Artwork Hierarchy and Structure

By default, the Illustrator hierarchy is structured such that the top elements are in front and the bottom elements are in back.

Move a layer, sublayer, or element

In the **Layers** panel, do the following (**Figure 6.12**):

- Click+drag the layer, sublayer, or element until a blue bar appears in the desired location and then release.

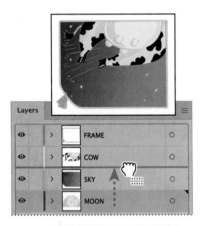

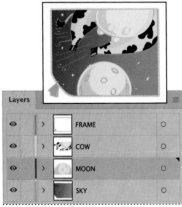

FIGURE 6.12 Moving a layer's position up in the Layers panel and the result

Nest a layer, sublayer, or element

In the **Layers** panel, do the following (**Figure 6.13**):

- Click+drag the layer, sublayer, or element onto the layer you want nest into, and then release the mouse when the destination layer highlights blue.

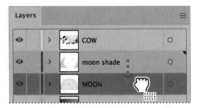

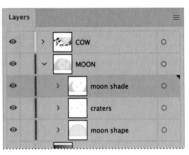

FIGURE 6.13 Nesting a layer inside another and the result

Delete a layer or sublayer

With the layer or sublayer active in the **Layers** panel, do either of the following:

- Click the **Delete Selection** button (**Figure 6.14**).
- Choose **Delete** [*element name*] from the panel menu.

FIGURE 6.14 Deleting an element using the Delete Selection button

Flatten layers

Flattening consolidates all your artwork's visible elements into a single layer.

In the **Layers** panel, do the following (**Figure 6.15**):

1. Select the layer you want the elements consolidated into.
2. Choose **Flatten Artwork** from the panel menu.

Merge elements

Merging layers or sublayers lets you select which elements you want to consolidate.

In the **Layers** panel, do the following (**Figure 6.16**):

1. Use the **Command/Ctrl** or **Shift** key to select the layers or sublayers you want to merge.
2. Choose **Merge Selected** from the panel menu.

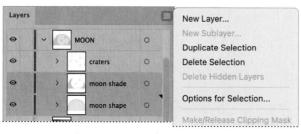

FIGURE 6.15 Consolidating all the artwork elements by selecting Flatten Artwork from the Layers panel menu

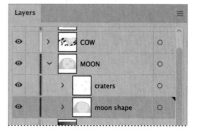

FIGURE 6.16 Consolidating selected sublayers by selecting Merge Selected from the Layers panel menu

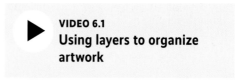

VIDEO 6.1
Using layers to organize artwork

Arrange objects using commands

The stacking order for objects can be changed within their sublayer or group by doing any of the following:

- Choose **Object** > **Arrange** > **Bring to Front** to move an object to the top position (**Figure 6.17**).

- Choose **Object** > **Arrange** > **Bring Forward** to move an object up one position.

- Choose **Object** > **Arrange** > **Send Backward** to move an object down one position.

- Choose **Object** > **Arrange** > **Send to Back** to move an object to the bottom position.

Arrange new objects using drawing modes

Drawing mode options let you choose whether you draw over, under, or inside existing elements in the same layer.

To select a drawing mode, choose any of the following from the lower section of the toolbar:

- Choose **Draw Normal** to place new objects on top of all the existing ones in the same layer (**Figure 6.18**).

- Choose **Draw Behind** to place new objects under all the existing ones in the same layer, or behind the currently selected element (**Figure 6.19**).

- Choose **Draw Inside** to place new objects inside the currently selected one (**Figure 6.20**).

> **TIP** Using the Draw Inside mode is similar to creating a clipping mask. To learn more, see Chapter 15.

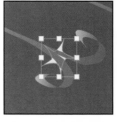

FIGURE 6.17 Moving an object to the top of its sublayer using the Bring to Front command

FIGURE 6.18 Adding an object in Draw Normal mode

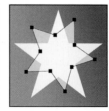

FIGURE 6.19 Adding an object in Draw Behind mode

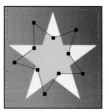

FIGURE 6.20 Adding an object in Draw Inside mode

Aligning and Distributing Elements

The **Align** panel (**Window** > **Align**) and menu commands let you neatly align and distribute objects.

TIP Align panel options are also available in the Control and Properties panel.

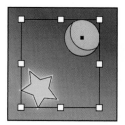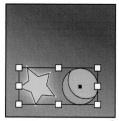

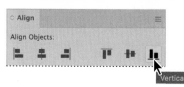

FIGURE 6.21 Vertically aligning selected objects using the Align panel

Align selected objects

With *two* or more objects selected, do either of the following:

- In the **Align Objects** section of the **Align** panel, click one or more alignment option (**Figure 6.21**).

- Choose **Object** > **Align** > [*menu option*]

Distribute selected objects

With *three* or more objects selected, do the following (**Figure 6.22**):

- In the **Distribute Objects** section of the **Align** panel, click one or more alignment option.

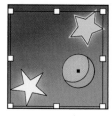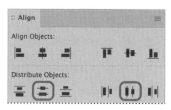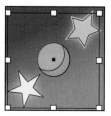

FIGURE 6.22 Vertically and then horizontally distributing selected objects

Use specific amounts to distribute selected objects

With *three* or more objects selected, do the following:

TIP If you don't see Distribute Spacing in the Align panel, choose Show Options from the panel menu.

1. In the **Distribute Spacing** section of the **Align** panel, enter the amount of space between the objects in the text boxes.

2. In the **Distribute Objects** section, click one or more alignment option.

Align or distributes objects using to a key object

By default, selected objects are aligned or distributed relative to their bounding box (**Figure 6.23**). To change this to a key object, do the following (**Figure 6.24**):

1. Select the objects to be aligned or distributed.

2. Click one of the selected object to use as a key object.

   Align to Key Object is automatically active in the **Align To** section of the **Align** panel, and a bold outline appears around the key object.

3. Click an option in the **Align Objects** or **Distribute Objects** section of the **Align** panel.

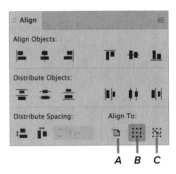

FIGURE 6.23
A. Align to Artboard
B. Align to Selection
C. Align to Key Object

TIP If you don't see the Align To section, choose Show Options from the Align panel menu.

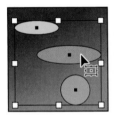
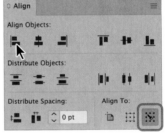
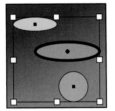
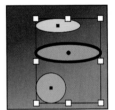

FIGURE 6.22 Aligning objects using a key object

Align or distributes objects to an artboard

Do the following:

1. Select the objects to be aligned or distributed.

2. Choose **Align to Artboard** in the **Align To** section of the **Align** panel.

3. Press **Shift** while clicking a blank area of the artboard you want to use.

4. Click an option in the **Align Objects** or **Distribute Objects** section of the **Align** panel.

VIDEO 6.2
Arranging and distributing objects

7

Selecting Elements

Adobe Illustrator provides several tools, modes, and commands for helping you select artwork elements easily and precisely.

In This Chapter

Selecting Objects Using Tools

Illustrator provides multiple tools for selecting elements (**Figure 7.1**).

Select objects and groups using the Selection tool

Do either of the following:

- Click an object or group (**Figure 7.2**).
- Click+drag a marquee over an object or group (**Figure 7.3**).

Add to a selection using the Selection tool

Do either of the following:

- Shift+click an unselected object or group (**Figure 7.4**).
- Shift+click and drag a marquee over an unselected object or group.

Subtract objects and groups from a selection

With the **Selection** tool active, do either of the following:

- Click a selected object or group.
- Click+drag a marquee over a selected object or group.

Deselect all selected elements

Do either of the following:

- Using the **Selection** tool, click a blank portion of the document or artboard.
- Choose **Select** > **Deselect**.

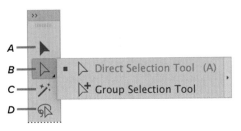

FIGURE 7.1
A. Selection tool **B.** Direct Selection tool
C. Magic Wand tool **D.** Lasso tool

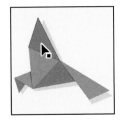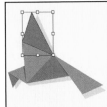

FIGURE 7.2 Selecting an object by clicking the Selection tool

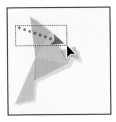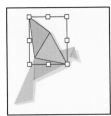

FIGURE 7.3 Selecting objects by dragging a marquee with the Selection tool

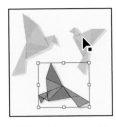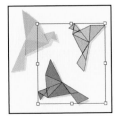

FIGURE 7.4 Shift-clicking an unselected group with the Selection tool to add it to the current selection

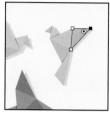

FIGURE 7.5 Clicking to select a point with the Direct Selection tool

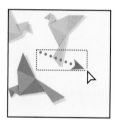
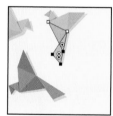

FIGURE 7.6 Selecting paths and points by dragging a marquee with the Direct Selection tool

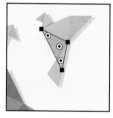

FIGURE 7.7 Clicking inside an object's path to select all of its paths and points with the Direct Selection tool

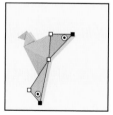

FIGURE 7.8 Shift-clicking an unselected point with the Direct Selection tool to add it to the current selection

Select paths or points using the Direct Selection tool

Do either of the following:

- Click a path or point (**Figure 7.5**).
- Click+drag a marquee over a path or point (**Figure 7.6**).

Select an object using the Direct Selection tool

Do the following:

- Click inside the object's path (**Figure 7.7**).

Add to a selected path or point using the Direct Selection tool

Do either of the following:

- Shift+click an unselected path or point (**Figure 7.8**).
- Shift+click and drag a marquee over an unselected object or group.

Subtract paths or points from selections

With the **Selection** tool active, do either of the following:

- Click a selected path or point.
- Click+drag a marquee over a selected object or group).

Deselect all paths or points

Do either of the following:

- Using the **Direct Selection** tool, click on a blank portion of the document or artboard.
- Choose **Select** > **Deselect**.

Select an object within a group using the Group Selection tool

With the **Group Selection** tool active, do the following (**Figure 7.9**):

- In the document window, click the object within the group.

Select an object's parent group using the Group Selection tool

With the **Group Selection** tool active, do the following (**Figure 7.10**):

- With the object selected in the document window, click the object a second time.

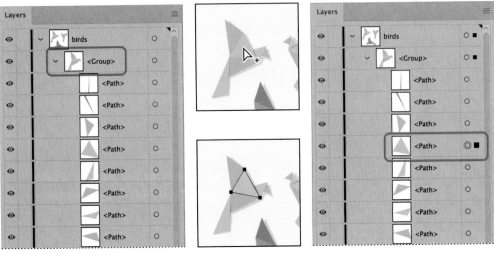

FIGURE 7.9 Selecting an object within a group using the Group Selection tool, and the result displayed in the Layers panel

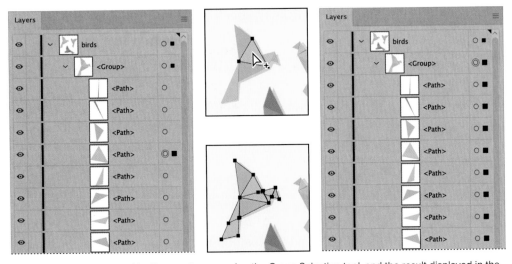

FIGURE 7.10 Selecting an object's parent group using the Group Selection tool, and the result displayed in the Layers panel

FIGURE 7.11 Selecting objects with the same fill using the Magic Wand tool

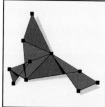

FIGURE 7.12 Adding to currently selected objects using the Magic Wand tool

FIGURE 7.13 Selecting an object using the Lasso tool

FIGURE 7.14 Selecting points and paths using the Lasso tool

Select objects assigned similar fill attributes using the Magic Wand tool

Do the following:

- Click the object assigned the fill (**Figure 7.11**).

Add to the current selection using the Magic Wand tool

Do the following:

- Shift+click an unselected object (**Figure 7.12**).

Subtract from the current selection using the Magic Wand tool

Do the following:

- Press **Alt/Option** and click a selected object.

Select objects using the Lasso tool

Do the following:

- Click+draw a freehand shape around the object (**Figure 7.13**).

Select points, paths, and objects using the Lasso tool

Do the following:

- Click+draw a freehand shape over the points and paths (**Figure 7.14**).

VIDEO 7.1
Working with selection tools

Working with Isolation Mode

Isolation modes lets you easily select and edit elements and layers by isolating them from the rest of the artwork.

While in isolation mode, all other artwork elements are dimmed and unselectable.

> **TIP** In Illustrator, elements can be paths, groups, raster images, meshes, and so on.

Isolate an element

Do any of the following:

- Double-click the element using the **Selection** tool.

- With an element or elements selected, click the **Isolate Selected Object** button in the **Control** panel (**Figure 7.15**).

- With an element or elements active in the **Layers** panel, choose **Enter Isolation Mode** from the panel menu.

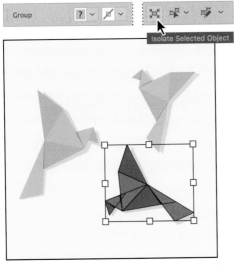
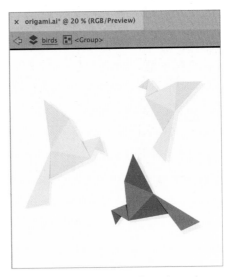

FIGURE 7.15 Activating isolation mode using the Isolate Selected Object button in the Control panel

Isolate a path within a group

Do the following:

1. Select the path with the **Direct Selection** tool or in the **Layers** panel.

2. In the **Control** panel, click the **Isolate Selected Object** button.

> **TIP** Continuing to double-click objects in Isolation mode drills down further into nested groups.

Isolate a layer or sublayer

Do the following:

1. Activate the layer in the **Layers** panel.

2. Choose **Enter Isolation Mode** from the panel menu.

> **TIP** To learn more about working with layers and groups, see Chapter 6.

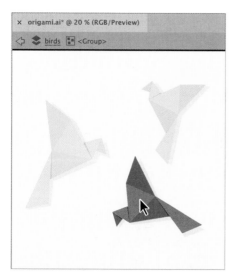

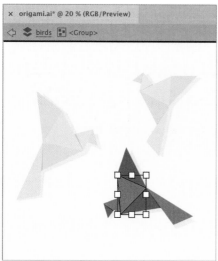

FIGURE 7.16 Selecting an element within an isolated group using the Selection tool

Select individual elements within an isolated group

With the group of elements in Isolation mode, do any of the following:

- Click the element using the **Selection** tool (**Figure 7.16**).
- Click a path or point using the **Direct Selection** tool.
- In the **Layers** panel, select the element.

Exit Isolation mode

Do any of the following:

- Press **Esc**.
- Click anywhere along the **Isolation Mode** bar.
- Double-click outside of the isolated group with the **Selection** tool.
- With the isolated elements deselected, click the **Back One Level** button in the **Control** panel.
- Click the **Exit Isolation Mode** button on the Isolation mode bar, one or more times.

TIP If a nested element has been isolated, the first click takes you back a level, and the second click exits Isolation mode.

Selecting Objects Using Commands

The **Select** menu provides commands to help you efficiently select all objects in your artwork, or specific objects based on their attributes and kind.

Select or deselect objects using the core Select commands

Do any of the following (**Figure 7.17**):

- Choose **Select** > **All** to select all the elements in your document.

- Choose **Select** > **All on Active Artboard** to select all the elements on the current artboard.

- Choose **Select** > **Deselect** to deselect all currently selected elements.

- Choose **Select** > **Reselect** to reactivate the selection of all elements that were deselected in the previous action.

- Choose **Select** > **Inverse** to select all unselected elements and deselect the currently selected ones (**Figure 7.18**).

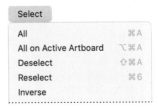

Select	
All	⌘A
All on Active Artboard	⌥⌘A
Deselect	⇧⌘A
Reselect	⌘6
Inverse	

FIGURE 7.17 Core Select menu commands

TIP Choosing **Select** > **Reselect** is particularly useful when you select multiple elements and then accidentally deselect them.

 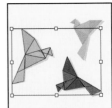

FIGURE 7.18 Choosing the Select > Inverse command to select all unselected elements and deselect the currently selected ones

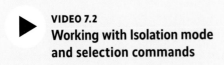

VIDEO 7.2
Working with Isolation mode and selection commands

Select objects with the same attributes

Do the following:

1. Select an object with the assigned attribute.

2. Choose **Select** > **Same** > [*command*] (**Figure 7.19**).

FIGURE 7.19 Choosing Select > Same menu commands

TIP To learn more about working with text and type, see Chapter 11.

Save selections

With the elements you want to save selected, do the following (**Figure 7.21**):

1. Choose **Select** > **Save Selection**.

2. Assign a name to the selection and click **Save**.

Select a saved selections

Do the following:

- Choose **Select** > [*saved selection name*} from the bottom of the menu.

Select all objects of the same kind

Do the following:

1. Deselect all elements.

2. Choose **Select** > **Object** > [*command*] (**Figure 7:20**).

TIP Elements need to be selected for the All on Same Layers command and Direction Handles command.

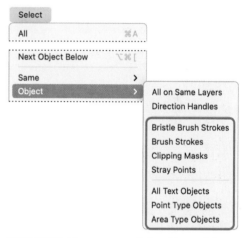

FIGURE 7.20 Menu commands for objects of the same kind

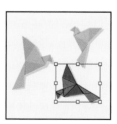

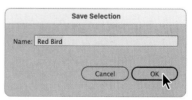

FIGURE 7.21 Saving a selection by choosing the Select > Save Selection command

Setting Selection Preferences

Selection & Anchor Display preferences let you adjust the tolerance for pixel selection and other options to suit your needs, especially when dealing with paths and points for complex elements.

Do the following (**Figure 7:22**):

1. Choose either:

   Illustrator > **Preferences** > **Selection & Anchor Display** (macOS)

   Edit > **Preferences** > **Selection & Anchor Display** (Windows)

2. Modify any of the following selection options:

   Tolerance to specify the pixel range for selecting anchor points. (Higher values increase the clickable area around an anchor point.)

   Snap to Point to snap objects to anchor points and guides. Specify the distance between the object and anchor point or guide when the snap occurs.

   Object Selection by Path Only to specify if filled objects are selectable by clicking anywhere within them.

   TIP Snap to Point also needs to be selected in the View menu to be active.

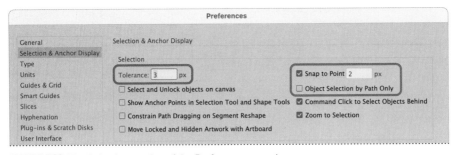

FIGURE 7.22 The Selection section of the Preferences panel

8

Customizing Strokes

Strokes can be much more than simple outlines for objects. Illustrator provides numerous tools for transforming them into visually rich elements.

Selecting a Stroke

An element's stroke (visible outline) can be selected using a number of methods.

Select an object's or the active stroke color using the toolbar

Do the following (**Figure 8.1**):

- Double-click the **stroke box** on the toolbar to open the Color Picker.

TIP To learn more about the Color Picker, see **Chapter 4**.

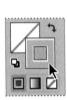
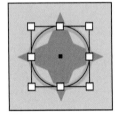
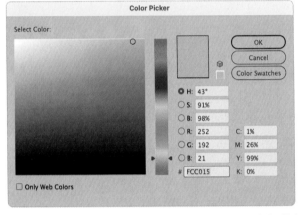

FIGURE 8.1 Selecting an object's stroke by double-clicking the stroke box in the toolbar to open the Color Picker

Select an object's stroke color or weight using panels

In the **Control**, **Properties**, or **Appearance** panel, do the following (**Figure 8.2**):

- Click the **stroke box** to open the Swatches panel and select a different color.

- Select or enter a new value in the **Stroke Weight** field.

TIP To learn more about the Swatches panel, see **Chapter 4**.

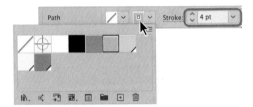

FIGURE 8.2 Accessing stroke color and weight options in the Control panel

TIP Shift-click the field to open the Color panel instead of the Swatches panel.

Working with the Stroke Panel

You can open the **Stroke** panel (**Figure 8.3**) from other panels or the Window menu.

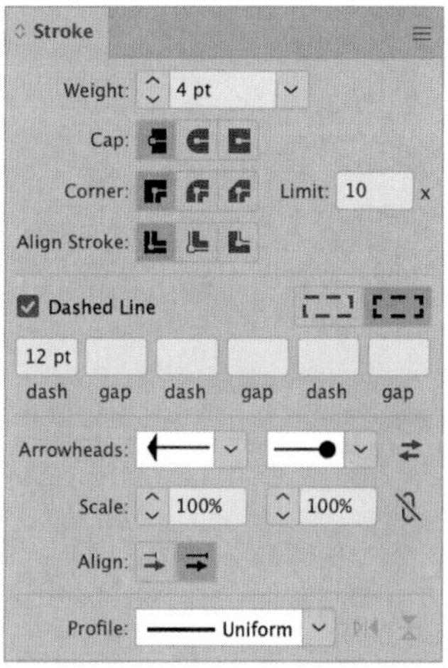

Open the Stroke panel independently

Do either of the following:

- Choose **Window** > **Stroke**, and then select **Show Options** from the panel menu.

- In the Essentials Classic workspace, click the **Stroke** panel thumbnail (**Figure 8.4**), and then select **Show Options** from the panel menu.

FIGURE 8.3 The Stroke panel

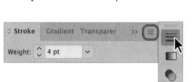

FIGURE 8.4 Accessing the Stroke panel from the thumbnail (panel menu highlighted)

Access the Stroke panel from another panel

In the **Control**, **Properties**, or **Appearance** panel, do the following (**Figure 8.5**):

- Click the word **Stroke** to open the panel.

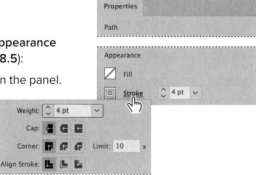

FIGURE 8.5 Accessing the Stroke panel from the Properties panel

Change the stroke weight

In **Weight** section of the **Stroke** panel, do any of the following:

- Click up or down arrow buttons to incrementally increase or decrease the weight.

- Enter a new amount in the value field.

- Click the pulldown menu button and select a new weight (**Figure 8.6**).

Change the stroke end cap

In the **Cap** section of the **Stroke** panel, do eany of the following:

- Select the **Butt Cap** (default) option to align the end of the stroke with the path.

- Select the **Round Cap** option to add an extended semicircular end (**Figure 8.7**).

- Select the **Projected Cap** option to add an extended rectangular end.

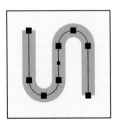  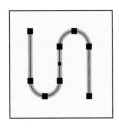

FIGURE 8.6 Decreasing an object's stroke weight using the menu

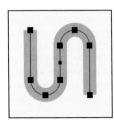 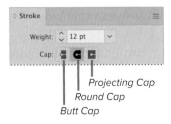 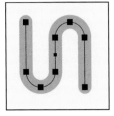

FIGURE 8.7 Object before and after applying a round end cap

Change the stroke corners

In the **Corner** section of the **Stroke** panel, do any of the following:

- Select the **Miter Join** (default) option to assign pointed corners.

- Select the **Round Join** option to assign elliptical corners (**Figure 8.8**).

- Select the **Bevel Join** option to assign squared corners.

> **TIP** Some tight corners may require increasing the Corner Limit to enable them to appear.

Change the stroke alignment

In the **Align Stroke** section of the **Stroke** panel, do any of the following:

- Select the **Align to Center** (default) option to position the stroke along the middle of the object's outline.

- Select the **Align to Inside** option to position the stroke within the object's outline (**Figure 8.9**).

- Select the **Align to Outside** option to position the stroke along the outer edge of the object's outline.

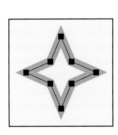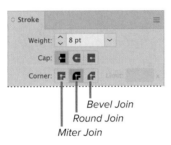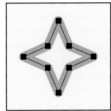

Bevel Join
Round Join
Miter Join

FIGURE 8.8 Object before and after applying a round join for the corners

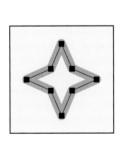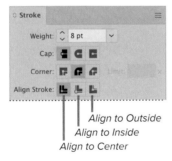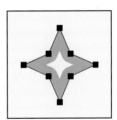

> **TIP** Inside and outside align options cannot be applied to open paths.

Align to Outside
Align to Inside
Align to Center

FIGURE 8.9 Object before and after applying an inside alignment to the stroke

Assign a dashed line

In the **Stroke** panel, do the following (Figure 8.10):

1. Select **Dashed Line**.

2. Specify the **Dash** segment length.

3. Specify the **Gap** space length.

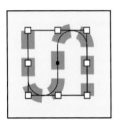

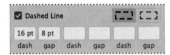

FIGURE 8.10 Dashed line properties applied to a path

Align Dashes to Corners and Path Ends

To position all the dashes at the corners of an object, maintaining visual consistency, do the following:

- Select the **Aligns Dashes to Corners and Path Ends, Adjusting Length to Fit** option (**Figure 8.12**).

Assign a dotted line

In the **Stroke** panel, do the following (Figure 8.11):

1. Select **Dashed Line**.

2. Select the **Round Cap** option.

3. Enter **0** for the **Dash** segment length.

4. Specify the **Gap** space length as appropriate.

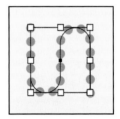

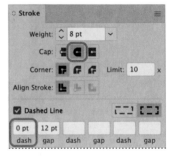

FIGURE 8.11 Dotted line properties applied to a path

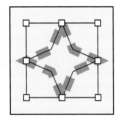

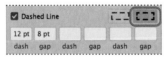

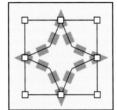

FIGURE 8.12 Object before and after applying dash alignment to the stroke

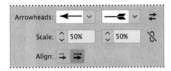

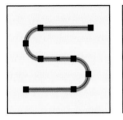

FIGURE 8.13 Path before and after applying and scaling arrowheads

 — *Swap Start and End Arrowheads*

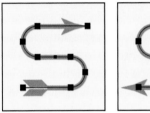

FIGURE 8.14 Arrowheads before and after swapping locations and extending beyond the path

Assign arrowheads to a path

In the **Stroke** panel, do either of the following (**Figure 8.13**):

- Select **Arrowheads** for the start and/or end points from the menus.

- Specify the appropriate **Scale** for the arrowheads.

TIP Arrowhead size correlates with stroke width.

Reposition arrowheads

In the **Stroke** panel, do either of the following (**Figure 8.14**):

- Click the **Swap Start and End Arrowheads** button to reverse the arrowheads.

- In the **Align** section, select whether to extend the arrowhead beyond the end of the path or position them at the tip.

TIP Arrowheads also swap positions when you revert the path direction.

Applying Varied Stroke Widths

Variable strokes let you stylize stroke widths and mimic traditional pen and brush strokes.

Assign a variable-width profile

In the **Stroke** or **Control** panel, do the following **(Figure 8.15)**:

- Select a **Profile** from the menu.

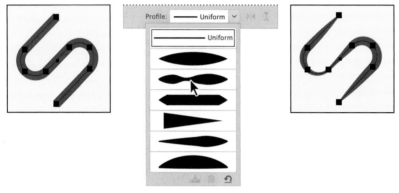

FIGURE 8.15 Path before and after applying a variable-width profile

Reverse an assigned variable-width profile

Do either the following **(Figure 8.16)**:

- Click the **Flip Along** button to vertically reverse the stroke.

- Click the **Flip Across** button to horizontally reverse the stroke.

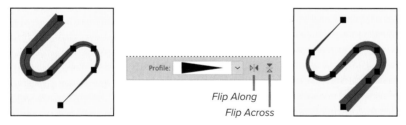

FIGURE 8.16 Path before and after clicking Flip Along button

FIGURE 8.17 The Width tool in the Essentials Classic toolbar

Vary stroke width using the Width tool

The **Width** tool (**Figure 8.17**) lets you create customized variable-width strokes.

With the **Width** tool active and the object you want to edit selected, do either of the following (**Figure 8.18**):

- Click+drag outward to add a wider stroke point.
- Click+drag inward to add a thinner stroke point.

FIGURE 8.18 Variably altering a path's stroke using the Width tool

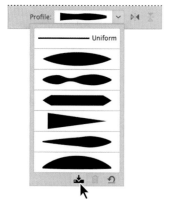

FIGURE 8.19 Adding a customized variable-width stroke to the profiles

Modify a variable-width stroke

With the **Width** tool active and the object you want to edit selected, do the following:

1. Double-click the stroked path on an already existing width point or at the position you want to adjust.
2. Adjust the settings in the **Width Point Edit** dialog box, and then click **OK**.

Save a variable-width stroke

To add a customized variable-width stroke as a profile, do the following (**Figure 8.19**):

1. Select the variable-width stroke object.
2. In the **Stroke** or **Control** panel, click the **Profile** menu.
3. At the bottom of the menu, click the **Add to Profiles** button.
4. In the **Variable Width Profile** dialog box, enter a **Profile Name,** and then click **OK**.

Converting Stroked Paths to Objects

Outlining a stroke lets you quickly convert it into a shape, which allows for additional flexibility and control when editing.

Create a shape from an object's stroke

Do the following (**Figure 8.20**):

1. Select the object with the stroke you want to convert.

2. Choose **Object** > **Path** > **Outline Stroke**.

3. (Optional) Choose **Object** > **Ungroup** to separate the new shape from the object's fill element. Then deselect both elements and select the new shape individually.

TIP Once you convert a stroke to a shape, it can be reverted only by choosing the Edit > Undo command, so make sure your artwork is in the its final form before converting the stroke.

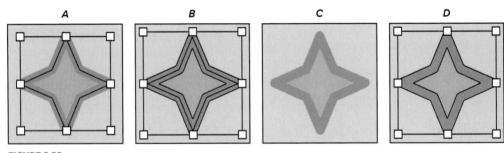

FIGURE 8.20
A. Object selected **B.** Stroke outlined **C.** Elements ungrouped and deselected **D.** New shape selected

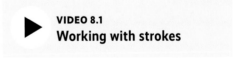

VIDEO 8.1
Working with strokes

Drawing Lines, Curves, and Paths

Pen and anchor point editing tools let you create paths using straight lines, curves, or a combination of both.

In This Chapter

Creating Lines and Curves with the Pen Tool

The **Pen** tool (**Figure 9.1**) lets you easily create straight lines, curves, and paths containing both lines and curves.

Draw a straight line

With the **Pen** tool active, do the following (**Figure 9.2**):

1. Position the cursor where you want the line to start.
2. Click to create the first anchor point.
3. Position the cursor where you want the line to end.
4. Click to create the second anchor point.

FIGURE 9.1 The Pen tool in the Essentials Classic toolbar

Draw a straight path

With the **Pen** tool active, do the following (**Figure 9.3**):

1. Click the point where you want the line to start.
2. Continue clicking to add anchor points for line segments.

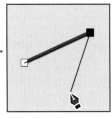 

FIGURE 9.3 Drawing a path by clicking multiple points with the Pen tool

TIP Pressing Shift as you click constrains the Pen tool to 45-degree increments.

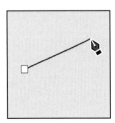 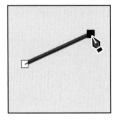

FIGURE 9.2 Drawing a straight line segment using the Pen tool

Draw an arc-curved path with the Pen tool

With the **Pen** tool active, do the following (Figure 9.4):

1. Position the cursor where you want the curve to begin.
2. Click+drag to set the slope of the curve.
3. Position the cursor where you want the curve to end.
4. Click+drag in the *opposite* direction of the previous slope direction.
5. Release the mouse.

Draw an S-curved path with the Pen tool

With the **Pen** tool active, do the following (Figure 9.5):

1. Position the cursor where you want the curve to begin.
2. Click+drag to set the slope of the curve.
3. Position the cursor where you want the curve to end.
4. Click+drag in the *same* direction of the previous slope direction.
5. Release the mouse.

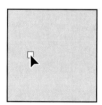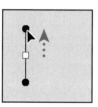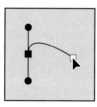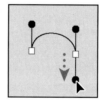

FIGURE 9.4 Drawing an arc-curve using the Pen tool

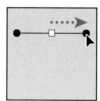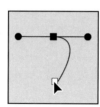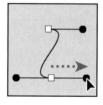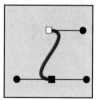

FIGURE 9.5 Drawing an S-curve using the Pen tool

TIP To learn about editing existing paths, see Chapter 14.

Draw a curve after a straight line

After creating a straight line segment with the Pen tool, do the following (**Figure 9.6**):

1. Position the Pen tool over the line endpoint until the Convert-point icon appears.

2. Click+drag to set the slope of the curve.

3. Position the cursor where you want the curve to end.

4. Click or click+drag to set the curve.

5. Release the mouse.

Draw a straight line after a curve

After creating the curve segment with the Pen tool, do the following (**Figure 9.7**):

1. Position the Pen tool over the curve endpoint until the Convert-point icon appears.

2. Click to set the line start point.

3. Position the cursor where you want the line to end.

4. Click to set the endpoint.

5. Release the mouse.

  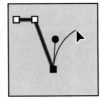 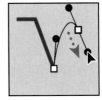 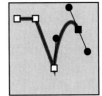

FIGURE 9.6 Drawing a curve after a straight line

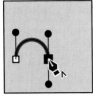  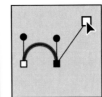 

FIGURE 9.7 Drawing a straight line after a curve

TIP You can also add a corner and second curve segment to an existing curve by click+dragging when adding the anchor points.

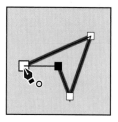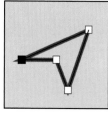

FIGURE 9.8 Clicking to close a straight path with the Pen tool

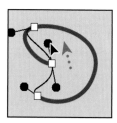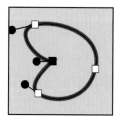

FIGURE 9.9 Clicking to close a curved path with the Pen tool

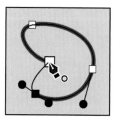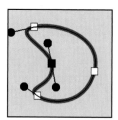

FIGURE 9.10 Click+dragging to close a curved path with the Pen tool

Close a straight path

With the Pen tool active, do the following (**Figure 9.8**):

1. Hover over the first anchor point until a small circle appears with the Pen tool icon.

2. Click to close the path.

Close a curved path

Hovering over the first anchor point until a small circle appears with the Pen tool icon, do either of the following:

- Click to close the path (**Figure 9.9**).
- Click+drag to close the path (**Figure 9.10**).

TIP **Pressing Option/Alt when closing the curve breaks the handle pairing of the closing anchor point.**

End an open path

Do any of the following:

- Press **Enter** or **Return**.
- Press **Esc**.
- Press **P**.
- Select a different tool.
- Choose **Select** > **Deselect**.
- **Ctrl-click** (Windows) or **Command-click** (macOS) on a blank space in the document.

Using the Curvature Tool

The Curvature tool (**Figure 9.11**) combines and simplifies many features of the Pen tool, letting you intuitively create paths. With each click, the Curvature tool adjusts the curve segments for the smoothest shape.

FIGURE 9.11 The Curvature tool in the Essentials Classic toolbar

Draw a smooth shape with the Curvature tool

With the Curvature tool active, do the following (**Figure 9.12**):

1. Click in two locations to add the first segment points.

2. Using the rubber band preview as a guide, add additional points. (See the "Rubber Band Preview" sidebar for more information.)

3. Close the object by clicking the starting anchor point.

Draw an open smooth path with the Curvature tool

With the Curvature tool active, do the following:

1. Click in two locations to add the first segment points.

2. Using the rubber band preview as a guide, add additional points.

3. Press **Esc** to end drawing the path.

FIGURE 9.12 Drawing a smooth shape by clicking to add points with the Curvature tool

Add a corner point with the Curvature tool

While creating segments with the Curvature tool, do the following (**Figure 9.13**):

- Double-click to add a point.

TIP Double-clicking existing curve points with the Curvature tool also converts them to corners.

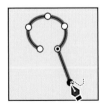

FIGURE 9.13 Drawing corner segments for a shape by double-clicking with the Curvature tool to add points

Rubber Band Preview

By default, the Pen tool and Curvature tool display a preview of the next path segment based on the current position of the cursor. This helps you accurately position the tool as you draw.

However, if you find it distracting, you can turn it off in the **Preferences** dialog box, under the **Selection & Anchor Display** section (**Figure 9.14**).

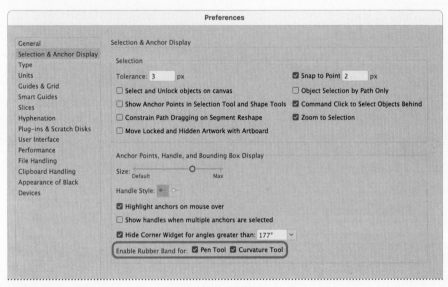

FIGURE 9.14 Rubber Band preview setting options in the Preferences dialog box

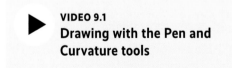

VIDEO 9.1
Drawing with the Pen and Curvature tools

Using the Line Segment Tools Group

The **Line Segment** tools group (**Figure 9.15**) lets you precisely create or manually draw single lines, arcs, spirals, and grids.

Create a line segment precisely

With the **Line Segment** tool active, do the following (**Figure 9.16**):

1. Position the cursor where you want the line segment to begin.

2. Click to open the **Line Segment Tool Options** dialog box.

3. Specify the line's **Length** and **Angle**.

4. (Optional) If you want to assign the current fill to the line, in addition to the stroke, select **Fill Line**.

5. Click **OK**.

> **TIP** Adding a fill to a line segment will be visible only if additional segments are added, and at different angles or curves.

Draw a line segment manually

With the **Line Segment** tool active, do the following (**Figure 9.17**):

- Click+drag to create the line segment.

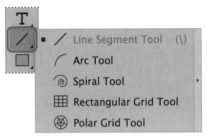

FIGURE 9.15 The Line Segment tools group in the Essentials Classic toolbar

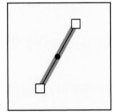

FIGURE 9.16 Precisely creating a line segment

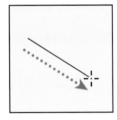

FIGURE 9.17 Manually drawing a line segment

Create an arc precisely

With the **Arc** tool active, do the following (**Figure 9.18**):

1. Position the cursor where you want the arc to begin.

2. Click to open the **Arc Segment Tool Options** dialog box.

3. Click a point on the **Reference Point Locator** to set the point of origin for the arc.

4. Enter the **Length X-Axis** (width) and **Length Y-Axis** (height).

5. Select the **Type** (Open or Closed).

6. Select the direction of the arc from **Base Along**.

   X Axis is horizontal.

   Y Axis is vertical.

7. Specify the direction of the arc's **Slope**.

   Concave (negative values) slope inward.

   Convex (positive values) slope outward.

8. (Optional) If you want to assign the current fill to the arc, in addition to the stroke, select **Fill Arc**.

9. Click **OK**.

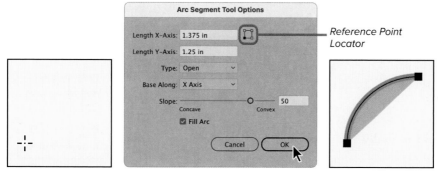

FIGURE 9.18 Precisely creating a filled arc

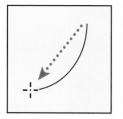

FIGURE 9.19 Manually drawing an arc

Draw an arc manually

With the **Arc** tool active, do the following (**Figure 9.19**):

- Click+drag to create the arc.

TIP Manually drawn Line Segment group elements use the currently assigned settings from the tool's Options dialog box.

Create a spiral precisely

With the **Spiral** tool active, do the following (**Figure 9.20**):

1. Position the cursor where you want the center of the spiral to be.

2. Click to open the **Spiral** dialog box.

3. Enter the size of the **Radius** (the distance from the center for the outermost segment).

4. Enter the amount of **Decay** (percentage of decrease for each spiral wind).

5. Enter the number of **Segments** (each full spiral wind has four segments).

6. Select a **Style** (direction of the spiral).

7. Click **OK**.

 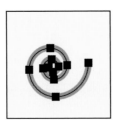

FIGURE 9.20 Precisely creating a spiral

Draw a spiral manually

With the **Spiral** tool active, do the following:

1. Position the cursor where you want the center of the spiral to be.

2. Click+drag to create the spiral.

Create a rectangular grid precisely

With the **Rectangular Grid** tool active, do any of the following and then click **OK** (**Figure 9.21**):

1. Position the cursor where you want the grid to begin.

2. Click to open the **Rectangular Grid Tool Options** dialog box.

3. Click a point on the **Reference Point Locator** to set the point of origin for the grid.

4. Enter the **Width** and **Height**.

5. Enter the number of **Horizontal** (row) and **Vertical** (column) **Dividers**.

6. Specify the **Skew** (how the dividers are positioned).

7. Select **Use Outside Rectangle As Frame** if you want to replace the outer segments with a separate rectangle object.

8. Select **Fill Grid** if you want to assign the current fill to the grid.

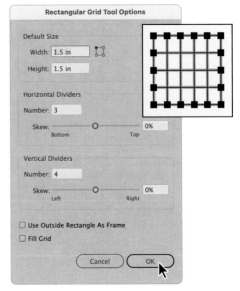

FIGURE 9.21 Precisely creating a rectangular grid

Create a polar grid precisely

With the **Polar Grid** tool active, do the following and then click **OK** (**Figure 9.22**):

1. Position the cursor where you want the grid to begin.

2. Click to open the **Polar Grid Tool Options** dialog box.

3. Click a point on the **Reference Point Locator** to set the point of origin for the grid.

4. Enter the **Width** and **Height**.

5. Enter the number of **Concentric** (row) and **Radial** (column) **Dividers**.

6. Specify the **Skew** (how the dividers are positioned).

7. Select **Create Compound Path from Ellipses** if you want to replace every other concentric circle with a compound path.

8. Select **Fill Grid** if you want to assign the current fill to the grid.

Draw a rectangular grid manually

With the **Rectangular Grid t**ool active, do the following:

1. Position the cursor where you want the center of the grid to be.

2. Click+drag to create the grid.

Draw a polar grid manually

With the **Polar Grid** tool active, do the following:

1. Position the cursor where you want the center of the spiral to be.

2. Click+drag to create the grid.

TIP Manually drawn Line Segment group elements use the currently assigned settings from the tool's Options dialog box.

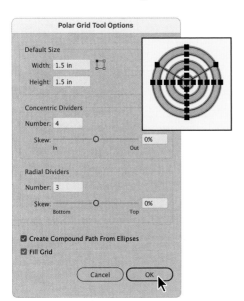

FIGURE 9.22 Precisely creating a polar grid

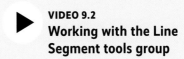

VIDEO 9.2
Working with the Line Segment tools group

Drawing with the Pencil Tool

The **Pencil** tool (**Figure 9.23**) lets you use traditional sketching strokes to create vector paths.

Draw a freehand open vector path

With the Pencil tool active, do the following (**Figure 9.24**):

1. Click+drag to draw the path.
2. Release the mouse to end the path.

Draw a freehand closed vector path

With the Pencil tool active, do the following (**Figure 9.25**):

1. Click+drag to draw the path, ending toward the beginning of the path.
2. Hover over the beginning of the path until a small circle appears with the Pencil tool icon.
3. Release the mouse to end the path.

Draw straight line segments

With the Pencil tool active, do the following (**Figure 9.26**):

1. Press the **Alt/Option** key.
2. Click+drag to draw the path.
3. Release the mouse to end the path.

TIP Pressing Shift+Alt/Option as you drag constrains the Pencil tool to 45-degree increments.

Draw additional line segments for a path

With the Pencil tool and the path active, do the following:

- Click the path start or end point, and continue drawing.

FIGURE 9.23 The Pencil tool (located under the Shaper tool) in the Essentials Classic toolbar

FIGURE 9.24 Drawing a freehand open vector path with the Pencil tool

FIGURE 9.25 Drawing a freehand open vector path with the Pencil tool

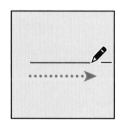
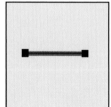

FIGURE 9.26 Drawing a straight path with the Pencil tool

Creating Shapes and Symbols

Illustrator provides a versatile set of tools for creating and customizing shapes, and using the symbol features helps you work efficiently and manage file size.

In This Chapter

Creating Rectangles and Squares

The **Rectangle** and **Rounded Rectangle** tools (**Figure 10.1**) let you easily add rectangular and square objects to your artwork.

Manually draw a rectangle

With the **Rectangle** tool active, do the following (**Figure 10.2**):

1. Position the cursor where you want the rectangle to begin.

2. Click+drag diagonally to set the size of the rectangle

3. Release the mouse.

> **TIP** Pressing Alt/Option as you click+drag sets the center of the rectangle as the point of origin.

Precisely create a rectangle

With the **Rectangle** tool active, do the following (**Figure 10.3**):

1. Position the cursor where you want the upper-left corner of the rectangle to be.

2. Click to open the **Rectangle** dialog box.

3. Enter dimensions for the **Width** and **Height**, and then click **OK**.

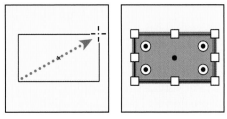

FIGURE 10.1 The Rectangle and Rounded Rectangle tools in the Essentials Classic toolbar

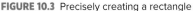

FIGURE 10.2 Manually drawing a rectangle

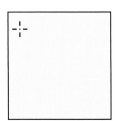
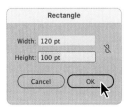

FIGURE 10.3 Precisely creating a rectangle

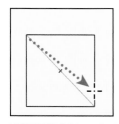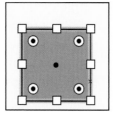

FIGURE 10.4 Manually drawing a square

FIGURE 10.5 Manually rounding the corners of a rectangle using the Corner Widget

TIP If you need the rectangle corner radius to be precise, create it by using the Rounded Rectangle tool or by entering the radius in the Shape properties of the Transform panel.

Manually draw a square

With the **Rectangle** tool active, do the following (**Figure 10.4**):

1. Position the cursor where you want the rectangle to begin.

2. Press **Shift** as you click+drag diagonally to set the size of the square.

3. Release the mouse.

Round the corners of a rectangular or polygon object

With the **Rectangle**, **Polygon**, or **Selection** tool active and the rectangular or polygon element selected, do the following (**Figure 10.5**):

1. Hover over the Corner Widget until the cursor displays an arc.

2. Click+drag inward to round the corners

TIP If the Corner Widgets are not visible, choose **View > Show Corner Widget.**

Manually draw a rounded rectangle or square

With the **Rounded Rectangle** tool active, do the following:

1. Position the cursor where you want the rectangle to begin.

2. Click+drag diagonally to set the size of the rounded rectangle

3. Release the mouse.

TIP Pressing Shift as you draw constrains the rounded rectangle to square proportions.

TIP Pressing the Up or Down Arrow keys as you draw increases or decreases the corner radius.

Precisely create a rounded rectangle or square

With the **Rounded Rectangle** tool active, do the following (**Figure 10.6**):

1. Position the cursor where you want the upper-left corner of the rounded rectangle to be.

2. Click to open the **Rounded Rectangle** dialog box.

3. Enter dimensions for the **Width, Height**, and **Corner Radius**.

4. Click **OK**.

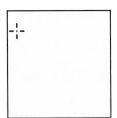 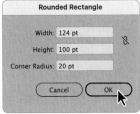 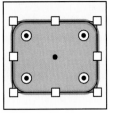

FIGURE 10.6 Precisely creating a rounded rectangle

Modify the default corner radius

To modify the default corner radius for rounded shapes, do the following:

1. Choose **Edit/Illustrator > Preferences > General**.

2. In the **General** section of the **Preferences** dialog box, enter a new **Corner Radius** (**Figure 10.7**).

3. Click **OK**.

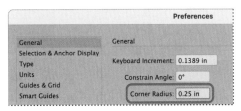

FIGURE 10.7 The Corner Radius setting in the Preferences dialog box

Creating Ovals, Circles, and Pie Segments

The **Ellipse** tool (**Figure 10.8**) lets you easily add oval objects, circular objects, and pie slice shapes.

Manually draw an oval or circle

With the **Ellipse** tool active, do the following (**Figure 10.9**):

1. Position the cursor where you want the ellipse to begin.

2. Click+drag diagonally to set the size of the ellipse.

3. Release the mouse.

> **TIP** Pressing Shift as you click+drag constrains the ellipse to a circle.

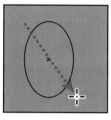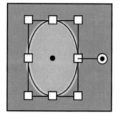

FIGURE 10.9 Manually drawing an oval

> **TIP** Pressing Alt/Option as you click+drag sets the center of the ellipse as the point of origin.

FIGURE 10.8 The Ellipse tool in the Essentials Classic toolbar

Precisely create an oval or circle

With the **Ellipse** tool active, do the following:

1. Position the cursor where you want the upper-left corner of the ellipse to be.

2. Click to open the **Ellipse** dialog box.

3. Enter dimensions for the **Width** and **Height**, and then click **OK**.

Create a pie shape from an oval or circle

With the **Ellipse** or **Selection** tool active and the elliptical element selected, do the following (**Figure 10.10**):

1. Hover over the arc widget until the cursor displays a pie shape.

2. Click+drag the arc widgets to add a pie slice into the ellipse or convert the ellipse to a slice shape.

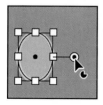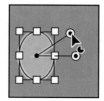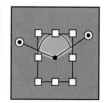

FIGURE 10.10 Creating pie segment shapes from an ellipse

Creating Polygons

The **Polygon** tool (**Figure 10.11**) lets you easily add shapes with straight sides and equal angles to your artwork.

Manually draw a polygon

With the **Polygon** tool active, do the following (**Figure 10.12**):

1. Position the cursor where you want the center of the polygon to be.

2. Click+drag diagonally to set the size of the polygon.

3. Release the mouse.

> **TIP** Pressing Shift as you click+drag constrains the angle of the polygon, allowing no rotation as you create it.

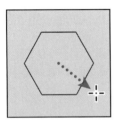

FIGURE 10.12 Manually drawing a polygon

> **TIP** The default number of sides for a manually drawn polygon in Illustrator is six.

FIGURE 10.11 The Polygon tool in the Essentials Classic toolbar

Precisely create a polygon

With the **Polygon** tool active, do the following:

1. Position the cursor where you want the center of the polygon to be.

2. Click to open the **Polygon** dialog box.

3. Enter the amount for the **Radius** and **Number of Sides**, and then click **OK**.

Modify the number of sides for an existing polygon

With the **Polygon** or **Selection** tool active and the polygon element selected, do the following (**Figure 10.13**):

1. Hover over the side number widget until the cursor displays a +/– icon.

2. Click+drag the widget up to decrease or down to increase the number of sides.

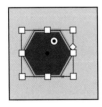
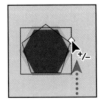
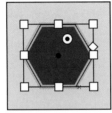
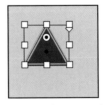

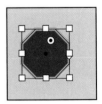

FIGURE 10.13 Decreasing and increasing the number of polygon sides

Creating Stars

The **Star** tool (**Figure 10.14**) lets you easily add customized star shapes to your artwork.

Manually draw a star

With the **Star** tool active, do the following (**Figure 10.15**):

1. Position the cursor where you want the center of the star to be.

2. Click+drag diagonally to set the size of the star.

3. Release the mouse.

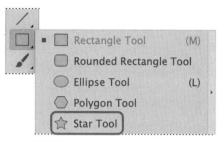

FIGURE 10.14 The Star tool in the Essentials Classic toolbar

FIGURE 10.15 Manually drawing a star

TIP Pressing Shift as you draw constrains the angle of the star angle of the star, allowing no rotation as you create it..

TIP Pressing the Command/Ctrl as you draw constrains the size of Radius 2.

TIP Pressing the Up or Down Arrow keys as you draw increases or decreases the number of points.

Precisely create a star

With the **Star** tool active, do the following (**Figure 10.16**):

1. Position the cursor where you want the center of the star to be.

2. Click to open the **Star** dialog box.

3. Enter the size for **Radius 1** (outer) and **Radius 2** (inner), and enter the number of **Points**.

4. Click **OK**.

TIP The larger radius will be used as the outer radius. If Radius 1 is smaller, then the star will be upside down, but you notice this only when it has an odd number of points.

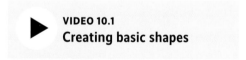

VIDEO 10.1
Creating basic shapes

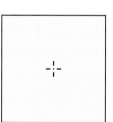

FIGURE 10.16 Precisely creating a star

Saving Objects as Symbols

If you need to reuse shapes and objects that you've created, saving them as symbols using the **Symbols** panel (**Window > Symbols**) can help you work efficiently and manage file size.

Create a symbol from an object

With the **Symbols** panel open, do the following:

1. Select the object(s).

2. Do any of the following:

   Click+drag the artwork onto the **Symbols** panel **(Figure 10.17)**.

   Click the **New Symbol** button at the bottom of the panel.

   Choose **New Symbol** from the panel menu.

3. Customize the symbol as needed, and click **OK** (Figure 10.18).

TIP Choosing Dynamic Symbol (default) allows you to override the appearance of a symbol instance while keeping the original symbol.

Use a symbol

In the **Symbols** panel, do the following:

- Drag the symbol onto the document window.

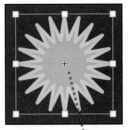
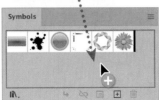

FIGURE 10.17 Click+dragging an object onto the Symbols panel

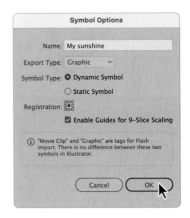

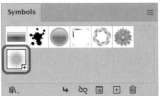

FIGURE 10.18 Adding a new symbol to the Symbols panel

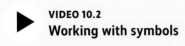
VIDEO 10.2
Working with symbols

Adding and Customizing Text

Illustrator provides powerful features for adding customized text to your artwork. Type can be transformed from simple text to eye-catching visual elements.

In This Chapter

Adding Text

Illustrator provides three methods for adding horizontal or vertical text to your artwork: *point*, *area* (text frames), and *type on a path* (**Figure 11.1**).

TIP Point text flows in one line from the point until you press Return or Enter. It also scales differently than area text.

Add point text

Do the following (**Figure 11.2**):

1. Select the **Type** or **Vertical Type** tool.

2. Click where you want the text to begin.

3. Enter the text.

4. Either deselect the text by clicking away from it or select the text by clicking the **Selection** tool.

FIGURE 11.2 Adding point text

TIP By default when using the Type tool or Vertical Type tool, Illustrator adds placeholder text before you beginning entering text.

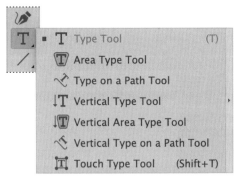

FIGURE 11.1 The Type tools group in the Essentials Classic toolbar

TIP To learn about the Touch Type tool, see "Formatting Character Settings" in this chapter.

Add area text

Do the following (**Figure 11.3**):

1. Select the **Type** or **Vertical Type** tool.

2. Click+drag diagonally to define the text boundaries.

3. Enter the text.

4. Either deselect the text by clicking away from it or select the text by clicking the **Selection** tool.

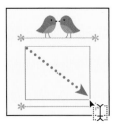

FIGURE 11.3 Adding area text

Adjust area text fitting

If the defined boundary is too small or overly large to accommodate the text, do either of the following:

- Adjust the dimensions by dragging the bounding box anchors.
- Double-click the middle handle at the bottom of bounding box to fit the length of the frame to the text (**Figure 11.4**).

FIGURE 11.4 Double-clicking the bottom area text handle to fit the frame to accommodate the text

TIP To learn more about text flow and accommodating overset text, see "Managing Text Content" in this chapter.

Convert area text to point text

Do the following (**Figure 11.5**):

- Double-click the middle handle on the right side of the bounding box.

FIGURE 11.5 Double-clicking an area text frame handle on the right to convert it to point text

Convert point text to area text

Do the following:

- Double-click the middle handle on the right side of the bounding box.

Add area text using a shape as the boundary

Do the following (**Figure 11.6**):

1. Select either the **Type**, **Vertical Type**, **Area Type**, or **Vertical Area Type** tool.
2. Click anywhere on the shape's edge.
3. Enter the text.
4. Either deselect the text by clicking away from it or select the text by clicking the **Selection** tool.

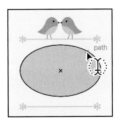

FIGURE 11.6 Adding area text using a shape as the boundary and the result

TIP To define a new line of point or area text, simply press Enter or Return.

Add text to an open path

Do the following (**Figure 11.7**):

1. Select either the **Type**, **Vertical Type**, **Type on a Path**, or **Vertical Type on a Path** tool.

2. Click on the path's edge where you want the text to start.

3. Enter the text.

4. Either deselect the text by clicking away from it or select the text by clicking the **Selection** tool.

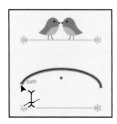 

FIGURE 11.7 Adding text to an open path

Add text to a closed path or shape

Do the following:

1. Select the **Type on a Path** or **Vertical Type on a Path** tool.

2. Click anywhere on the shape's edge.

3. Enter the text.

4. Either deselect the text by clicking away from it or select the text by clicking the **Selection** tool.

Manually adjust the type position along the path

Do the following (**Figure 11.8**):

1. Select the text using the **Selection** tool.

2. Click+drag either the middle bracket or one of the end brackets along the path.

FIGURE 11.8 Adjusting the text path by dragging the middle bracket

Flip the type on path position

Do the following (**Figure 11.9**):

1. Select the text object using the **Selection** tool.

2. Click+drag the middle bracket across the path.

FIGURE 11.9 Flipping the type on path position

Apply a path attribute to the text

With the text path selected, do the following:

- Choose **Type** > **Type on a Path** > [*attribute name*] (**Figure 11.10**).

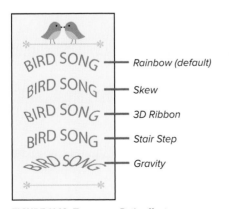

FIGURE 11.10 Type on a Path effects

Adjust the type on path vertical alignment

With the text path selected, do the following:

1. Either double-click the **Type on a Path** tool or choose **Type** > **Type on Path** > **Type on a Path Options**.

2. In the **Type on a Path Options** dialog box, select an **Align to Path** option (**Figure 11.11**)

3. Click **OK** to apply the change (**Figure 11.12**).

FIGURE 11.11 Selecting an Align to Path option

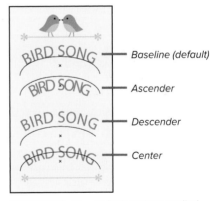

FIGURE 11.12 Align to Path options applied

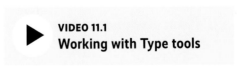

VIDEO 11.1
Working with Type tools

Choosing Fonts

Illustrator allows you to easily select and preview your available fonts and provides quick access for downloading additional ones.

Access the Character panel

Font and other typographic settings are available in the **Character** panel. To access the panel, do any of the following:

- Choose **Window** > **Type** > **Character** to open the panel independently (**Figure 11.13**).

- With text selected, click the word **Character** in the **Control** panel (**Figure 11.14**).

- With text selected, in the **Properties** panel under **Character**, click the **More Options** button (**Figure 11.15**).

FIGURE 11.13 The Character panel

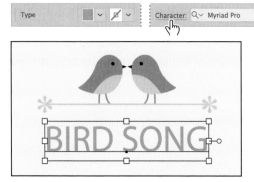

FIGURE 11.14 Clicking the word Character in the Control panel to access the Character panel

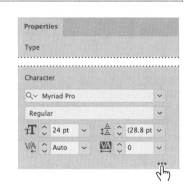

FIGURE 11.15 Clicking the More Options button in the Properties panel to access the Character panel

TIP Fonts can also be chosen using **Type** > **Font** > *[font name]*.

About Font Types

Illustrator supports several types of fonts:

 OpenType is a format created by Adobe and Microsoft for both macOS and Windows platforms.

 Variable is an OpenType format that lets you flexibly adjust attributes such as the weight, width, and slant.

 SVG is an OpenType format designed for glyphs and emojis, allowing single characters to have multiple colors and gradients.

 Adobe Fonts are available with a Creative Cloud subscription. The library contains thousands of fonts.

 Type 1 fonts were developed by Adobe and have since been widely replaced by OpenType fonts.

 TrueType was developed by Apple and is a cross-platform format for both macOS and Windows platforms.

 Multiple Master fonts were developed by Adobe and have since been widely replaced by OpenType variable fonts.

 Composite fonts were developed for East Asian languages and have been widely replaced by OpenType fonts.

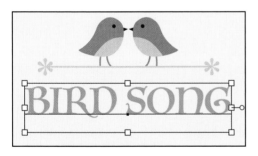

Select a font family

With the text or text object selected, do the following (**Figure 11.16**):

1. In either the **Character**, **Control**, or **Properties** panel, click the **Font Family** pulldown menu button.

2. Select a new font from the menu.

TIP If you know the name of the font, you can also type it in the Font Family field.

TIP The font families displayed depend on which fonts you have on your system and which filters you've applied.

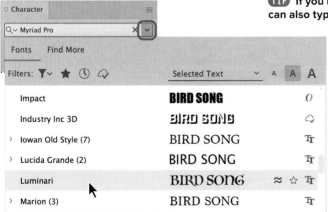

FIGURE 11.16 Selecting a font from the Font Family menu

Narrow font search options

In the **Filters** section of the **Font Family** menu, do any of the following (**Figure 11.17**):

- Click the **Filter Fonts by Classification** button and then select the classification and property options you want to include (**Figure 11.18**).

- Click the **Show Favorite Fonts** button to display only the fonts you've selected as favorites.

  TIP See Figure 11.20 for an example of adding a font to favorites.

- Click the **Show Recently Added** button to limit the number of fonts displayed to only the most recently added (by default, ten).

- Click the **Show Activated Fonts** button to display only fonts activated from the Adobe Fonts library.

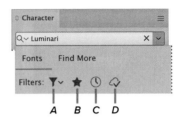

FIGURE 11.17
A. Filter Fonts by Classification
B. Show Favorite Fonts
C. Show Recently Added
D. Show Activated Fonts

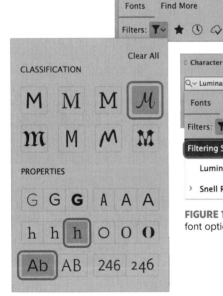

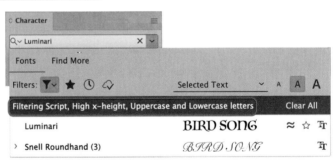

FIGURE 11.18 Selecting classification and properties options to filter the font options (left) and the result (above)

Find similar fonts

Do the following (**Figure 11.19**):

- Hover over a font and then click the **Show Similar Fonts** icon.

Add font to favorites

Do the following (**Figure 11.20**):

- Hover over the selected font and then click the **Add to Favorites** icon.

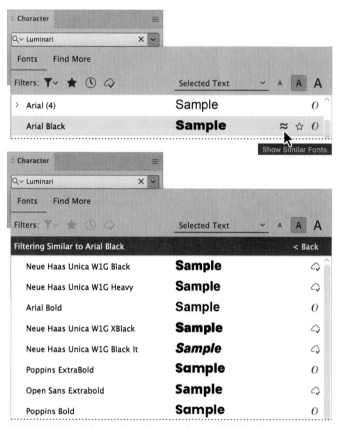

FIGURE 11.19 Clicking to show fonts similar to Arial Black

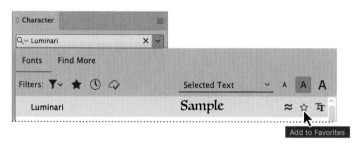

FIGURE 11.20 Clicking to add Luminari font to favorites

Apply a font style

With the text or text object selected,
do the following (**Figure 11.21**):

1. In either the **Character**, **Control**, or
   Properties panel, click the **Set the Font
   Style** menu button.

2. Select a new style from the pulldown
   menu.

FIGURE 11.21 Selecting a font style from the menu

Modify variable fonts

With the variable font text or text object
selected, do the following (**Figure 11.22**):

1. In either the **Character**, **Control**, or
   Properties panel, click the **Variable
   Font** button.

2. Either use the sliders or enter new
   values to adjust the settings

TIP To learn more about variable fonts,
see the "About Font Types" sidebar in this
chapter.

FIGURE 11.22 Adjusting the variable font value settings

Customizing Character Settings

Illustrator provides numerous options for customizing text character settings.

Adjust the font size

To adjust the size of the text, do the following:

1. Select the text using either the **Type** or **Selection** tool.

2. In either the **Character**, **Control**, or **Properties** panel, adjust the size by clicking the **Font Size** (**A** in **Figure 11.23**) menu button and selecting a new size or entering a new value in the field.

> **TIP** Font size can also be changed by choosing **Type > Size > [font size]**.

Adjust the kerning

To adjust the spacing between two characters, do the following (**Figure 11.24**):

1. With the **Type** tool active, click between the two characters.

2. In either the **Character**, **Control**, or **Properties** panel, adjust the distance by clicking the **Kerning** (**B** in **Figure 11.23**) menu button and selecting a new distance or entering a new distance in the field.

FIGURE 11.24 Clicking between two characters using the Text tool and the result after increasing the kerning to 200

> **TIP** The Selection tool selects the text object. The Type tool can select a single character or a range.

FIGURE 11.23
A. Font Size **B.** Kerning **C.** Leading **D.** Tracking

Adjust the leading

To adjust the spacing between lines of text, do the following:

1. Select the text using either the **Type** or **Selection** tool.

2. In either the **Character**, **Control**, or **Properties** panel, adjust the spacing by clicking the **Leading** (**C** in **Figure 11.23**) menu button and selecting a new spacing or entering a new value in the field.

Adjust the tracking

To adjust the spacing between selected-characters, do the following:

1. Select the text using either the **Type** or **Selection** tool.

2. In either the **Character**, **Control**, or **Properties** panel, adjust the spacing by clicking the **Tracking** (**D** in **Figure 11.23**) menu button and selecting a new spacing or entering a new value in the field.

Adjust the scale

To adjust the vertical or horizontal scale of the characters using the **Character** panel (**A** and **C** in **Figure 11.25**), do the following (**Figure 11.26**):

1. Select the text using either the **Type** or **Selection** tool.

2. In the **Character** panel, adjust the scale by clicking either the **Vertical Scale** or **Horizontal Scale** menu button and selecting a new percentage or entering a new percentage in the field.

Adjust the baseline shift

To move characters up or down relative to the surrounding text baseline using the **Character** panel (**B** in **Figure 11.25**), do the following (**Figure 11.27**):

1. Select the characters using the **Type** tool.

2. In the **Character** panel, adjust the position by clicking the **Baseline Shift** menu button and selecting a new amount or entering a new amount in the field.

TIP Positive numbers shift the characters above the baseline, and negative numbers shift the characters below the baseline.

Change the rotation of a character

To individually rotate selected characters using the **Character** panel (**D** in **Figure 11.25**), do the following (**Figure 11.28**):

1. Select the characters using the **Type** tool.

2. In the **Character** panel, adjust the rotation by clicking the **Character Rotation** menu button and selecting a new amount or entering a new amount in the field.

TIP Negative numbers rotate the characters clockwise, and positive numbers rotate the characters counterclockwise.

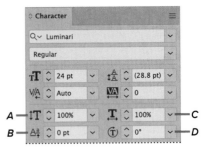

FIGURE 11.25
A. Vertical Scale **B.** Baseline Shift
C. Horizontal Scale **D.** Character Rotation

TIP If all the character setting options are not visible, choose Show Options from the panel menu to display them.

FIGURE 11.26 Increasing the horizontal scale of a selected character

FIGURE 11.27 Shifting a selected character above the baseline

FIGURE 11.28 Rotating a selected character counterclockwise

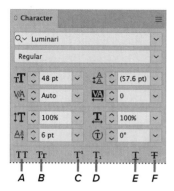

FIGURE 11.29
A. All Caps **B.** Small Caps
C. Superscript **D.** Subscript
E. Underline **F.** Strikethrough

Apply text treatments

To apply a text treatment to selected characters using the **Character** panel (**Figure 11.29**), do any of the following (**Figure 11.30**):

- Select **All Caps** to capitalize every selected character.

- Select **Small Caps** to assign small capitalization to selected lowercase characters.

- Select **Superscript** to shift the baseline up and decrease the scale of the selected characters.

- Select **Subscript** to shift the baseline down and decrease the scale of the selected characters.

- Select **Underline** to apply a line below the selected characters.

- Select **Strikethrough** to apply a line through the selected characters.

All Caps   *Small Caps*

Superscript   *Subscript*

Underline   *Strikethrough*

FIGURE 11.30 Applying text treatments

> **TIP** Scaling, superscript, and subscript baseline shift percentages can be modified in File > Document Setup > Type.

Using the Touch Type tool

To isolate and modify individual characters using the **Touch Type** tool, do the following (**Figure 11.31**):

1. Select the **Touch Type** tool (inside the **Type** tool group in Essentials Classic).

2. Click the character you want to modify and then use the bounding boxes to reposition, scale, and/or rotate the character.

FIGURE 11.31 Transforming a character using the Touch Type tool

Customizing Paragraph Settings

Access the Paragraph panel

Alignment, justification, distance between paragraphs, and indentation settings are available in the **Paragraph** panel (**Figure 11.32**). To access the panel, do either of the following:

- Choose **Window** > **Type** > **Paragraph** to open the panel independently.

- With text selected, click the word **Paragraph** in the **Control** panel (**Figure 11.33**).

- With text selected, in the **Properties** panel under **Paragraph**, click the **More Options** button (**Figure 11.34**).

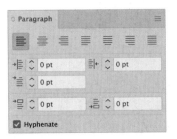

FIGURE 11.32 The Paragraph panel

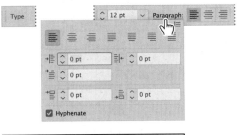

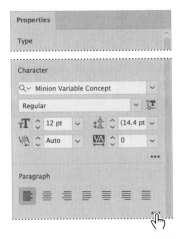

FIGURE 11.34 Clicking the More Options button in the Properties panel to access the Paragraph panel

FIGURE 11.33 Clicking the word Paragraph in the Control panel to access the Paragraph panel for a selected type object

FIGURE 11.35 Align options highlighted with Align Center selected

FIGURE 11.36 Justify options highlighted with Justify with Last Line Aligned Center selected

Adjust the alignment

To set the horizontal positioning of the selected paragraph lines, do the following (**Figure 11.35**):

- Select an **Align** option in the **Paragraph** panel.

Adjust the justification

Justified paragraphs align to both the left and right edges of the text frame. To determine how the last line is positioned, do either of the following:

- Select a **Justify** option in the **Paragraph** panel to set the alignment (left, center, right, or full) (**Figure 11.36**).

- To more specifically set justification options, in the **Paragraph** panel menu, choose **Justification** to open the dialog box and customize the settings (**Figure 11.37**).

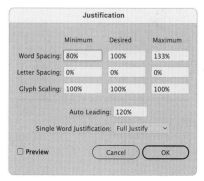

FIGURE 11.37 The Justification dialog box

Apply indentations using the Paragraph panel

To modify the amount of space between the selected paragraph and its vertical boundaries, do either of the following:

- Select or enter a new value in the **Left Indent** or **Right Indent** settings to indent the entire paragraph (**Figure 11.38**).

- Select or enter a new value in the **First-Line Left Indent** settings to indent only the first line of the paragraph (**Figure 11.39**).

Determine the space between paragraphs

With the text selected, do either of the following:

- In the **Paragraph** panel, select or enter a new value in the **Space Before** option.

- In the **Paragraph** panel, select or enter a new value in the **Space After** option (**Figure 11.40**).

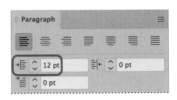

FIGURE 11.38 Setting the left paragraph text indent

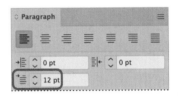

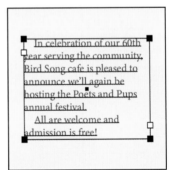

FIGURE 11.39 Setting the first-line paragraph text indent

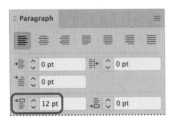

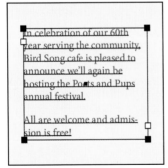

FIGURE 11.40 Setting the Space Before amount

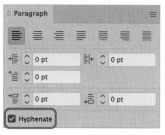

FIGURE 11.41 The Language option menu in the Character panel

TIP If the language menu is not visible in the Character panel, choose Show Options from the panel menu to display it.

FIGURE 11.42 The Hyphenate option selected in the Paragraph panel

Set hyphenation options

To determine how lines and words break for selected text, do any of the following:

- Choose a language in the **Character** panel to determine how words are divided (**Figure 11.41**).

- Select or deselect the **Hyphenate** option in the **Paragraph** panel (**Figure 11.42**).

- To more specifically set the hyphenation options, choose **Hyphenation** from the **Paragraph** panel menu to open the dialog box and customize the settings (**Figure 11.43**).

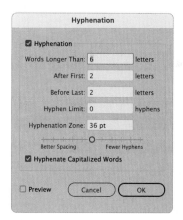

FIGURE 11.43 The Hyphenation dialog box

Adjust area type positioning

To specify the vertical position of the selected area type within its frame, do the following:

- Select an option in the **Area Type** section of either the **Control** or **Properties** panel (**Figure 11.44**).

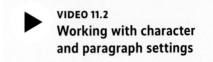

VIDEO 11.2
Working with character and paragraph settings

  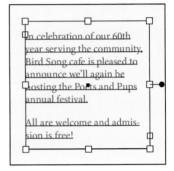

FIGURE 11.44 Changing an area type's vertical positioning and the result

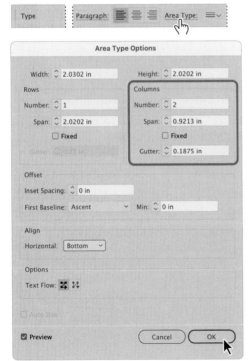

Modify area type options

The **Area Type Options** dialog box provides additional formatting options for selected text, such as number of columns and gutter size. To open the dialog box, do either of the following (**Figure 11.45**):

- Click the words **Area Type** in the **Control** panel.

- In the **Area Type** section of the **Properties** panel, click **More Options**.

FIGURE 11.45 Opening the Area Type Options dialog box and increasing the number of columns

Working with Tabs

The **Tabs** panel (**Figure 11.46**) lets you set indents and add tab stops.

> **TIP** To access the Tabs panel, choose **Window > Type > Tabs.**

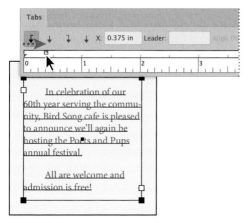

FIGURE 11.46
A. Left-Justified tab **B.** Center-Justified tab
C. Right-Justified tab **D.** Decimal-Justified tab
E. Left Indent marker **F.** First-Line Indent marker
G. Position Panel Above Text button

Align the Tabs panel to text

With the text selected, do the following:

- Click the **Position Panel Above Text** button (**G** in **Figure 11.46**).

Apply first-line indentations

With the text selected, do either of the following:

- Drag the **First-Line Indent** marker to the right (**Figure 11.47**).

- Select the **First-Line Indent** marker and enter a positive value in the **X** field.

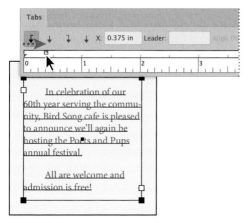

FIGURE 11.47 Dragging the First-Line Indent marker to indent the first line of the selected paragraphs

Apply hanging indentations

With the text selected, do either of the following:

- Drag the **Left Indent** marker to the right (**Figure 11.48**).

- Select the **Left Indent** marker and enter a positive value in the **X** field.

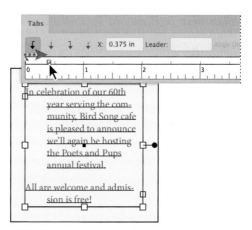

FIGURE 11.48 Dragging the Left Indent marker to indent the selected paragraphs but not the first line

> **TIP** If you need the Tabs panel to display a longer ruler, click+drag the lower corner of the panel to resize it.

Apply tab stops to text

Do any of the following (**Figure 11.49**):

- Using the **Type** tool, click the text where you want to insert the tab stop and press **Tab**.

- In the **Tabs** panel, select a tab justification button.

- In the **Tabs** panel, either click along the ruler or enter an **X** value to add the tab stop.

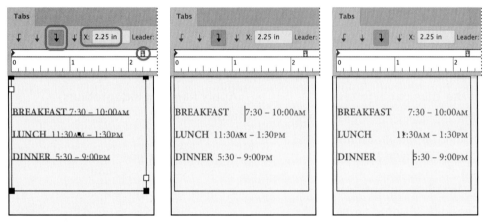

FIGURE 11.49 Adding a Right-Justified tab to the selected text and inserting tab stops in the lines

Apply leaders to tab stops

With the text and tab stop selected, do the following (**Figure 11.50**):

- In the **Leader** field, enter one to eight characters and then press **Enter** or **Return**.

TIP Periods and spaces are common leader characters.

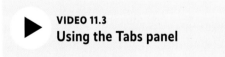
VIDEO 11.3
Using the Tabs panel

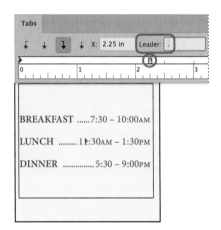

FIGURE 11.50 Adding a leader to a tab stop

Using Character and Paragraph Styles

If you repeatedly use certain format attributes, the **Character Styles** and **Paragraph Styles** panels let you save text attributes to help maintain consistency and work efficiently.

TIP To access the Character Styles or Paragraph Styles panel, choose **Window** > **Type** > **Character Styles or Paragraph Styles**. By default, they are grouped together.

Create a character or paragraph style

With the formatted characters selected, do either of the following:

- Click the **New Style** button to add the new style using the default name.

- Choose **New Character Style** or **New Paragraph Style** from the panel menu then enter a name in the dialog box and click **OK** (Figure 11.51).

Apply a character or paragraph style

Do the following (**Figure 11.52**):

1. Using the **Type** tool, select the characters or paragraph for applying the style.

2. In the **Character Styles** or **Paragraph Styles** panel, click the style.

TIP If you use the New Style button, you can rename the new style by double-clicking the name.

 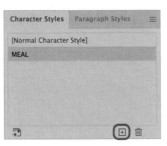

FIGURE 11.51 Creating a new character style from selected text

 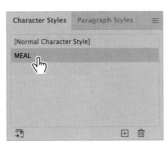

FIGURE 11.52 Applying a character style to selected text

Modify a character or paragraph style

Do the following (**Figure 11.53**):

1. In the **Character Styles** or **Paragraph Styles** panel, select the style.

2. Double-click the style you want to modify.

3. Customize the settings as needed and then click **OK**.

TIP Styles can also be customized in the New Character Style and New Paragraph Style dialog boxes. When you modify a style, the formatting of any text with that style applied is updated.

FIGURE 11.53 Modifying a character style's character fill color

Working with Special Characters

Illustrator provides several tools for adding special characters to your text.

TIP Glyph options are font-dependent.

Insert a glyph

Do the following (**Figure 11.54**):

1. Using the **Type** tool, click the text where you want to insert the glyph.

2. In the **Glyphs** panel, double-click a glyph to insert it.

FIGURE 11.54 Inserting a glyph using the Glyphs panel

Replace an individual character using the Glyphs panel

Do the following (**Figure 11.56**):

1. Using the **Type** tool, select the character.

2. In the **Glyphs** panel, double-click a glyph to replace the selected character.

TIP To access the Glyphs panel, choose **Window > Type > Glyphs.**

Replace an individual character with glyphs using the in-context menu

Do the following (**Figure 11.55**):

1. Using the **Type** tool, select the character.

2. Select an alternative glyph from the in-context menu that appears automatically.

FIGURE 11.55 Selecting an alternative glyph character using the on-canvas menu

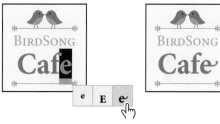

FIGURE 11.56 Replacing a character using the Glyphs panel

Applying OpenType formats

Do the following:

1. Using the **Type** or **Selection** tool, select the text assigned the OpenType font you want to customize.

2. In the **OpenType** panel (**Figure 11.57**), choose the rules you want to apply to the glyphs.

Insert special, white space, and break characters using the Type menu

Do the following:

1. Using the **Type** tool, click the text where you want to insert the character.

2. From the **Type** menu, choose one of the following:

   Insert Special Character to select symbols, hyphens, dashes, and quotation marks (**Figure 11.58**).

   Insert WhiteSpace Character to select specifically proportioned space characters (**Figure 11.59**).

   Insert Break Character > Forced Line Break to insert a new line without beginning a new paragraph.

TIP To access the OpenType panel, choose **Window > Type > OpenType**.

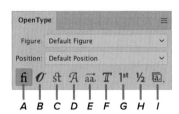

FIGURE 11.57
A. Standard Ligatures **B.** Contextual Alternates **C.** Discretionary Ligatures **D.** Swash **E.** Stylistic Alternate **F.** Titling Alternates **G.** Ordinals **H.** Fractions **I.** Stylistic Sets

FIGURE 11.58 Choosing Insert Special Character from the Type menu

FIGURE 11.59 Choosing Insert WhiteSpace Character from the Type menu

Managing Text Content

Illustrator provides numerous tools for managing text content, including importing text, finding words, and accommodating text flow.

Import text as a new file

Do the following (**Figure 11.60**):

1. Choose **File** > **Open**.
2. In the **Open File** dialog box, select the text document and click **Open**.
3. (Optional) If you are opening a Word document, select the settings you want to customize and then click **OK**.

Import text into an existing file

Do the following (**Figure 11.61**):

1. Choose **File** > **Place**.
2. In the **Place File** dialog box, select the text document and click **Open**.
3. (Optional) If you are opening a Word document, select the settings you want to customize and then click **OK**.

Export text as text document

Do the following:

1. Using the **Type** tool, select the text.
2. Choose **File** > **Export As**.
3. For **Format**, choose **Text Format (TXT)**.
4. Enter a name for the file.
5. Click **Export** (macOS) or **Save** (Windows).
6. In the dialog box, select the **Platform** and **Encoding** methods; then click **Export** (**Figure 11.62**).

TIP Illustrator supports importing most Microsoft Word (.doc and .docx) formats, as well as Rich Text Format (.rtf) and plain text (.txt).

FIGURE 11.60 A Microsoft Word document opened in Illustrator

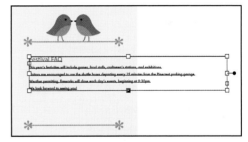

FIGURE 11.61 A Microsoft Word document placed into an existing Illustrator file

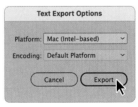

FIGURE 11.62 Selecting the platform and encoding methods for exported text

Search for text

To search for and replace text, do the following (**Figure 11.63**):

1. Choose **Edit** > **Find and Replace**.

2. In the dialog box, enter the text to search for in the **Find** field.

3. Click the **Find** button to search for and select the text.

FIGURE 11.63 Searching for a word

Replace text

Do the following (**Figure 11.64**):

1. In the **Find and Replace** dialog box, enter the replacement text in the **Replace** field.

2. (Optional) Refine the search and replacement parameters by selecting any of the appropriate options.

3. Click any of the following:

   Replace to replace only the selected found word.

   Replace & Find to replace the found word and continue searching.

   Replace All to replace all instances of the found word in the file.

4. Click **Done**.

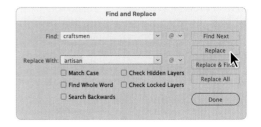

FIGURE 11.64 Replacing a word and the result

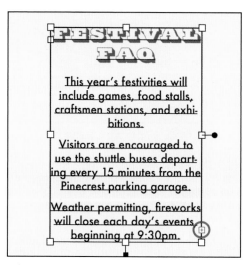

FIGURE 11.65 Example of overset text

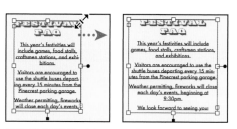

FIGURE 11.66 Resizing a text frame to accommodate overset text

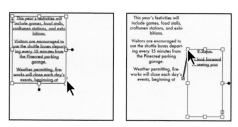

FIGURE 11.67 Adding a threaded text frame to accommodate overset text

Resize overset text frames

Overset refers to text that does not fit inside its boundaries. This is indicated by a red box with a plus sign in the lower-right corner (**Figure 11.65**).

To resize an overset text frame, do either of the following:

- Drag the bounding box anchors to resize the text (**Figure 11.66**).
- Double-click the middle handle at the bottom of the bounding box to fit the length of the frame to the text.

Thread text

Threading refers to adding linked frames to accommodate the extra text.

To thread overset text, do the following (**Figure 11.67**):

1. Double-click the overset text icon.

2. Click or click+drag to add the threaded text frame.

Convert type to outlines

Converting text to paths (outlines) is helpful when sharing a file with someone who will not have the required fonts or when making edits to the text shapes.

To convert selected text to compound paths, do the following (**Figure 11.68**):

1. Choose **Type** > **Create Outlines**.

2. (Optional) To edit the paths, release groups and compound paths by choosing the following:

   Object > **Ungroup**

   Object > **Compound Paths** > **Release**

FIGURE 11.68 Converting text to paths

TIP Once you convert your type to outlines, you will lose the ability to make any text editing or formatting changes, so make sure you have done all those modifications beforehand.

Working with Brushes and Sketch Tools

Illustrator provides a variety of freehand drawing tools that allow you to replicate traditional drawing techniques and create objects using freehand methods.

In This Chapter

Working with the Brushes Panel

The **Brushes** panel (**Figure 12.1**) provides brush stroke stylization options as you paint with the **Paintbrush** tool. The brushes can also be applied to existing paths.

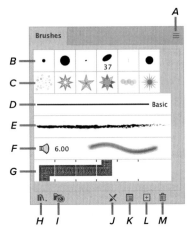

FIGURE 12.1
A. Panel menu **B.** Calligraphic brushes
C. Scatter brushes **D.** Default stroke
(removes brush stroke from the path)
E. Art brush **F.** Bristle brush
G. Pattern brush **H.** Brush Libraries menu
I. Libraries panel **J.** Remove Brush Stroke
K. Options of Selected Object **L.** New Brush
M. Delete Brush (removes the brush from
the file and the panel)

TIP The brushes displayed in the Brushes panel are document-dependent.

Brush Types

The different types of brushes that Illustrator provides achieve very different effects:

Calligraphic brushes replicate calligraphy pen strokes and are centered along the path.

Scatter brushes randomly disperse copies of an object along the path.

Art brushes stretch a shape evenly along the length of the path.

Bristle brushes replicate traditional bristle brush strokes.

Pattern brushes place a repeating tiled pattern along the path.

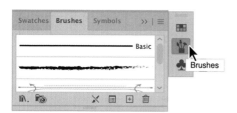

FIGURE 12.2 Accessing the Brushes panel from the Control panel

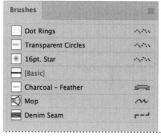

FIGURE 12.3 Accessing the Brushes panel by clicking the thumbnail

Access the Brushes panel

Do any of the following:

- Choose **Window** > **Brushes**.

- In the **Control** panel, click the **Brush Definition** field or pulldown menu (**Figure 12.2**).

- In the **Essentials Classic** workspace, click the **Brushes** thumbnail (**Figure 12.3**).

Manage the Brushes panel display options

From the **Brushes** panel menu (**Figure 12.4**), do any of the following:

- Select or deselect **Show** *[brush type name]* to display or hide brush types.

- Select **Thumbnail View** to display the brushes graphically or **List View** to see more detailed information.

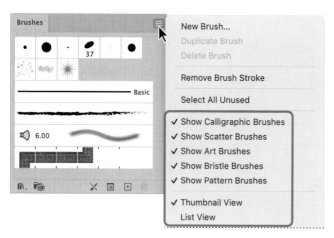

FIGURE 12.4 The Brushes panel menu's display options and the panel after deselecting Show Calligraphic Brushes and choosing List View

Apply a brush to an existing path

Do the following (**Figure 12.5**):

1. Select the object or object's path.

2. In the **Brushes** panel, click the brush you want to apply.

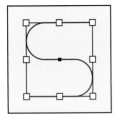 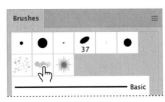 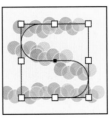

FIGURE 12.5 Applying a scatter brush to a selected object's path

Remove a brush stroke from a path

With the object or object's path selected, do any of the following in the **Brushes** panel:

- Click the **Remove Brush Stroke** button (**Figure 12.6**).

- Choose **Remove Brush Stroke** from the panel menu.

- Select the **Basic** option.

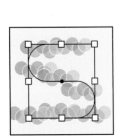 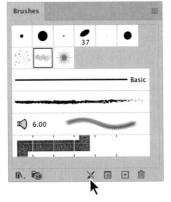 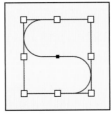

FIGURE 12.6 Removing a scatter brush from a selected object's path

Make a copy of a brush

In the **Brushes** panel, do the following (**Figure 12.7**):

1. Click the brush you want to duplicate.

2. Select **Duplicate Brush** from the panel menu.

TIP It's often a good idea to make a copy of a brush before making any changes to it.

Modify a brush

Do the following (**Figure 12.8**):

1. In the **Brushes** panel, Double-click the brush you want to change to open the brush type's dialog box.

2. Rename the brush and/or adjust the settings.

3. Click **OK**.

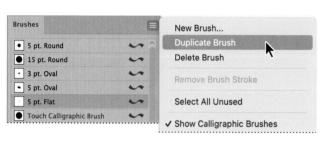
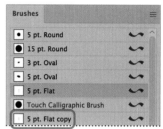

FIGURE 12.7 Making a copy of a calligraphic brush

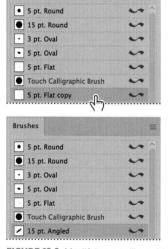
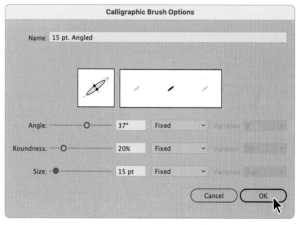

FIGURE 12.8 Modifying a calligraphic brush

TIP To learn more about brush type options, see "Creating Brushes" in this chapter.

VIDEO 12.1
Working with the Brushes panel

Using Brush Libraries

Brush libraries (**Figure 12.9**) provide a wide selection of preset brush collections that are included with Illustrator.

Access a brush library

Do any of the following:

- Choose **Window** > **Brush Libraries** and select a library from the submenus.

- In the **Brushes** panel, click the **Brush Libraries** menu and select a library from the submenus (**Figure 12.10**).

- In a brush library panel, click the **Brush Libraries** menu and select a library from the submenus.

- In a brush library panel, click either the **Load Previous Brush Library** or **Load Next Brush Library** button.

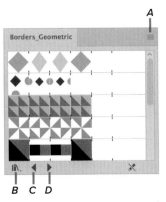

FIGURE 12.9
A. Panel menu **B.** Brush Libraries menu
C. Load Previous Brush Library
D. Load Next Brush Library

Add library brushes to the Brushes panel

In the library brush panel, do any of the following:

- Click the brush.

- Shift-click to select multiple brushes and then choose **Add to Brushes** from the panel menu.

- Shift-click to select multiple brushes and then drag them onto the **Brushes** panel.

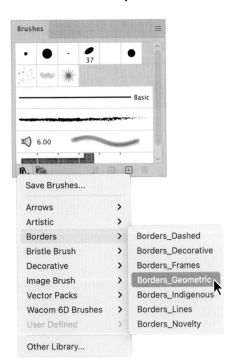

FIGURE 12.10 Selecting a brush library

Finding brush types in the libraries

Some brush libraries use different names than those in the **Brushes** panel. Here is where to find the brush type you're looking for:

Calligraphic brushes

- Artistic > Calligraphic
- Wacom 6D Brushes > 6D Art Pen Brushes

Scatter brushes

- Arrows > Standard
- Artistic > Artistic Ink
- Decorative > Decorative Scatter
- Decorative > Elegant Curl & Floral Brush Set
- Image Brush > Image Brush Library
- Wacom 6D Brushes > 6D Art Pen Brushes

Art brushes

- Arrows > Special
- Arrows > Standard
- Artistic > (all except Artistic Calligraphic)
- Decorative > Banners and Seals
- Decorative > Text Dividers
- Decorative > Elegant Curl & Floral Brush Set
- Image Brush > Image Brush Library
- Vector Packs > (all)

Pattern

- Borders > (all)
- Arrows > Pattern Arrows
- Decorative > Elegant Curl & Floral Brush Set
- Image Brush > Image Brush Library

Apply a library brush to a path

Do the following (**Figure 12.11**):

- With an object or object path selected, click the library brush to apply it to the **Brushes** panel.

> **TIP** Applying a library brush to a path automatically adds it to the Brushes panel.

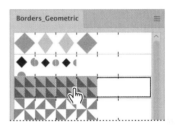

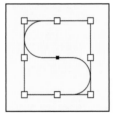 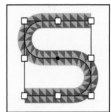

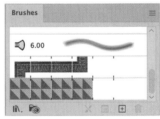

FIGURE 12.11 Applying a library brush to a path and adding it to the Brushes panel

Creating Brushes

You can create your own brushes for any brush type and save collections of brushes as libraries.

Create a calligraphic brush

Do the following (**Figure 12.12**):

1. In the **Brushes** panel, either click the **New Brush** button or choose **New Brush** from the panel menu.

2. In the **New Brush** dialog box, select **Calligraphic Brush** and then click **OK**.

3. In the **Calligraphy Brush Options** dialog box, enter a **Name** for the new brush.

4. Adjust the brush settings as needed, and then click **OK** to add the new brush to the **Brushes** panel.

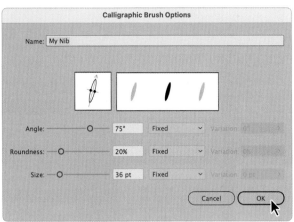

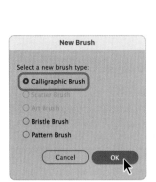
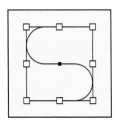
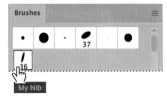
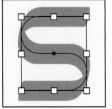

FIGURE 12.12 Creating and applying a new calligraphy brush

The Calligraphic Brush Options dialog box settings

These are the settings you can apply when creating or modifying a calligraphic brush:

- **Angle** sets the brush's rotation.
- **Roundness** sets whether the brush is circular or oval shaped. A setting of 100% applies a circular brush, and lower percentages create an oval-shaped brush.
- **Diameter** sets the length of the brush.

With the **Angle**, **Roundness**, and **Diameter** settings are the following additional options:

- **Fixed** means the brush has no variation to the setting.
- **Random** allows for random variations to the setting once you begin to draw.
- **Variation** determines how much change occurs as varying pressure is applied.

If you are using a tablet or stylus, the following options may also be available:

- **Pressure** is most useful for determining the *diameter* by varying the setting based on the pressure of a drawing stylus if you are using a graphics tablet.
- **Stylus Wheel** varies the *diameter* and other options to which it is applied if you are using an airbrush pen with a stylus wheel on its barrel and a graphics tablet that can detect that pen.
- **Tilt** is most useful for setting the *roundness* by varying the settings based on the angle of a drawing stylus if you are using a graphics tablet that can detect how close to vertical the pen is.
- **Bearing** is most useful for setting the *angle* if you are using a graphics tablet that can detect what direction the pen is tilted.
- **Rotation** is most useful for setting the *angle* of flat brushes if you are using a graphics tablet that can detect the rotation of the pen between your fingers.

TIP To modify an existing calligraphic brush, double-click the brush in the Brushes panel to open the Calligraphy Brush Options dialog box.

Create a scatter brush

Do the following (**Figure 12.13**):

1. Select the objects for the new brush.

2. In the **Brushes** panel, either click the **New Brush** button or choose **New Brush** from the panel menu.

3. In the **New Brush** dialog box, select **Scatter Brush** and then click **OK**.

4. In the **Scatter Brush Options** dialog box, enter a name for the new brush.

5. Adjust the brush settings as needed, and then click **OK** to add the new brush to the **Brushes** panel.

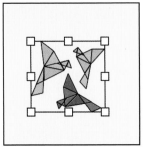

Selected objects for the brush

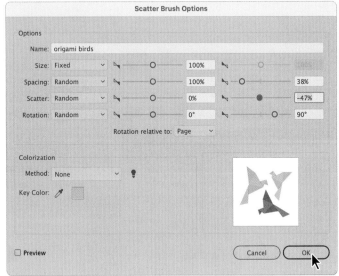

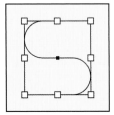

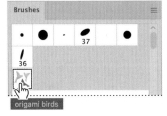

FIGURE 12.13 Creating and applying a new scatter brush

The Scatter Brush Options dialog box settings

These are the settings you can apply when creating or modifying a scatter brush:

- **Size** sets the diameter of the brush.
- **Spacing** sets the distance between artwork elements.
- **Scatter** sets how near the objects follow the path (negative values to the right, positive values to the left).
- **Rotation** sets the angle of the objects.
- **Rotation Relative To** sets whether the angle is relative to the page or the path.

With the **Size**, **Spacing**, **Scatter**, and **Rotation** settings are the following additional options:

- **Fixed** means the brush has no variation to the setting.
- **Random** allows for random variations to the setting.
- **Variation** specifies the range for **Random** if the setting is selected.

If you are using a tablet or stylus, the following additional options may also be available:

- **Pressure** is most useful for determining the *size* and *density* by varying the setting based on the pressure of a drawing stylus if you are using a graphics tablet.
- **Stylus Wheel** varies the *size* and *density* if you are using an airbrush pen with a stylus wheel on its barrel and a graphics tablet that can detect that pen.
- **Bearing** is most useful for setting the *rotation* if you are using a graphics tablet that can detect what direction the pen is tilted.
- **Rotation** is most useful for setting the *rotation* of brushes if you are using a graphics tablet that can detect the rotation of the pen between your fingers.

The **Colorization** settings let you set the color method for the brush using the following options:

- **None** keeps the colors the same as the original objects selected for the brush.
- **Tints** uses the active stroke color to replace black elements, and other colored elements convert to variations of the stroke color. White elements will remain the same.
- **Tints and Shades** uses the tints and shades of the active stroke color for the brush stroke. 50% gray is replaced by the stroke color. 100% black and white remain the same.
- **Hue Shift** converts the **Key Color** (by default, the prominent color in the artwork) to the active stroke color and rotates all the other brush colors in the color wheel proportionally to that. It's like using the color wheel in the **Recolor Artwork** feature when you move all the linked colors of the artwork to a different color.

TIP To modify an existing scatter brush, double-click the brush in the Brushes panel to open the Scatter Brush Options dialog box.

Create an art brush

Do the following (**Figure 12.14**):

1. Select the objects for the new brush.

2. In the **Brushes** panel, either click the **New Brush** button or choose **New Brush** from the panel menu.

3. In the **New Brush** dialog box, select **Art Brush** and then click **OK**.

4. In the **Art Brush Options** dialog box, enter a name for the new brush.

5. Adjust the brush settings as needed, and then click **OK** to add the new brush to the **Brushes** panel.

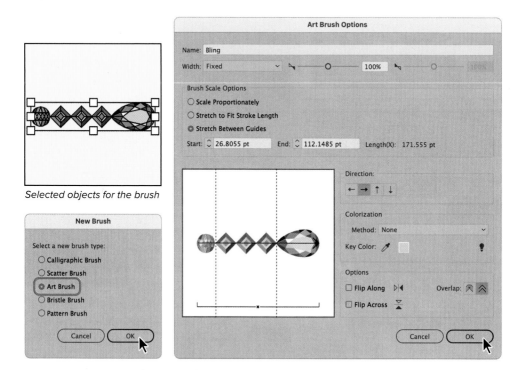

Selected objects for the brush

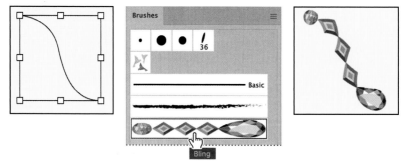

FIGURE 12.14 Creating and applying a new art brush

The Art Brush Options dialog box settings

These are the settings you can apply when creating or modifying an art brush:

Width modifies how wide the brush is relative to the original objects. Width settings are *fixed* unless you are using a tablet or stylus, in which case the following other options may also be available:

- **Pressure** is most useful for determining the width by varying the setting based on the pressure of a drawing stylus if you are using a graphics tablet.

- **Stylus Wheel** varies the width if you are using an airbrush pen with a stylus wheel on its barrel and a graphics tablet that can detect that pen.

- **Tilt** is most useful for setting the width by varying the settings based on the angle of a drawing stylus if you are using a graphics tablet that can detect how close to vertical the pen is.

- **Bearing** is most useful for setting the *rotation* if you are using a graphics tablet that can detect what direction the pen is tilted.

- **Rotation** is most useful for setting the *rotation* of brushes if you are using a graphics tablet that can detect the rotation of the pen between your fingers.

Brush Scale Options determine whether the brush is scaled proportionally, stretched to fit the length of the path, or if segments are stretched between specified guides.

Direction determines the direction in which the brush design is applied to the path (across or along) as well as the direction.

The **Colorization** settings let you set the color method for the brush using the following options:

- **None** keeps the colors the same as the original objects selected for the brush.

- **Tints** uses the active stroke color to replace black elements, and other colored elements convert to variations of the stroke color. White elements will remain the same.

- **Tints and Shades** uses the tints and shades of the active stroke color for the brush stroke. 50% gray is replaced by the stroke color. 100% black and white remain the same.

- **Hue Shift** converts the **Key Color** (by default, the prominent color in the artwork) to the active stroke color and rotates all the other brush colors in the color wheel proportionally to that. It's like using the color wheel in the **Recolor Artwork** feature when you move all the linked colors of the artwork to a different color.

Flip options change the orientation of the brush relative to the path.

Overlap options let you set whether to adjust corners to avoid overlaps and folds.

TIP To modify an existing art brush, double-click the brush in the Brushes panel to open the Art Brush Options dialog box.

Create a bristle brush

Do the following (**Figure 12.15**):

1. In the **Brushes** panel, either click the **New Brush** button or choose **New Brush** from the panel menu.

2. In the **New Brush** dialog box, select **Bristle Brush** and then click **OK**.

3. In the **Bristle Brush Options** dialog box, enter a name for the new brush.

4. Adjust the brush settings as needed, and then click **OK** to add the new brush to the **Brushes** panel.

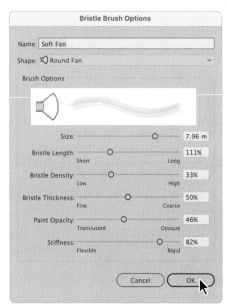

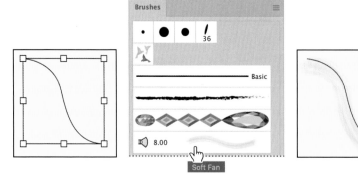

FIGURE 12.15 Creating and applying a new bristle brush

The Bristle Brush Options dialog box settings

These are the settings you can apply when creating or modifying a bristle brush:

Shape menu options provide 10 traditional brush model presets.

Size sets the diameter of the brush.

Bristle Length sets how long the bristles are from the tip to where they would meet the brush handle.

Bristle Density sets how many bristles are in a defined area based on the brush size and bristle length.

Bristle Thickness determines how coarse or fine the bristles are. The lower the number, the finer the bristles.

Paint Opacity determines the transparency of the paint being applied. The lower the number, the more transparent.

Stiffness determines the hardness of the bristles. The lower the number, the softer the bristles.

TIP **To modify an existing bristle brush, double-click the brush in the Brushes panel to open the Bristle Brush Options dialog box.**

Create a pattern brush

Do the following (**Figure 12.16**):

1. In the **Swatches** panel, add the pattern swatches to use for the brush.

2. In the **Brushes** panel, either click the **New Brush** button or choose **New Brush** from the panel menu.

3. In the **New Brush** dialog box, select **Pattern Brush** and then click **OK**.

4. In the **Pattern Brush Options** dialog box, enter a name for the new brush.

5. Adjust the brush settings as needed, and then click **OK** to add the new brush to the **Brushes** panel.

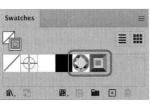

Pattern swatches for the brush

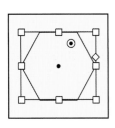

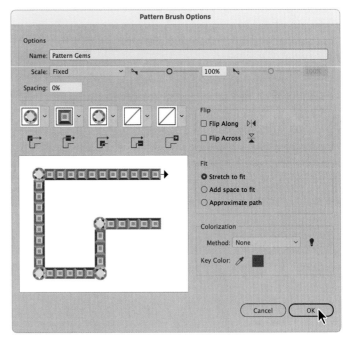

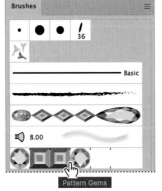

FIGURE 12.16 Creating and applying a new pattern brush

The Pattern Brush Options dialog box settings

These are the settings you can apply when creating or modifying a pattern brush:

Scale sets the size of the tiles relative to the original objects. Scale settings are *fixed* unless you are using a tablet or stylus, in which case the following options may also be available:

- **Pressure** is most useful for determining the *scale* by varying the setting based on the pressure of a drawing stylus if you are using a graphics tablet.

- **Stylus Wheel** varies the *scale* if you are using an airbrush pen with a stylus wheel on its barrel and a graphics tablet that can detect that pen.

- **Tilt** is most useful for setting the *scale* by varying the settings based on the angle of a drawing stylus if you are using a graphics tablet that can detect how close to vertical the pen is.

- **Bearing** is most useful for setting the *rotation* if you are using a graphics tablet that can detect what direction the pen is tilted.

- **Rotation** is most useful for setting the *rotation* of brushes if you are using a graphics tablet that can detect the rotation of the pen between your fingers.

Spacing sets the distance between the tiles.

Tile Buttons let you access the patterns in the Swatches panel to apply different tiles to different parts of the path.

Flip options mirrors the brush stroke (which consists of a sequence of tiles) along or across the path.

Fit options determine whether the tiles are stretched or compressed to fit the length of the path, if space is added to fit the path, or if the tiles are fitted approximately to accommodate the path length.

The **Colorization** settings let you set the color method for the tiles using the following options:

- **None** keeps the colors the same as the original objects selected for the brush.

- **Tints** uses the active stroke color to replace black elements, and other colored elements convert to variations of the stroke color. White elements will remain the same.

- **Tints and Shades** uses the tints and shades of the active stroke color for the brush stroke. 50% gray is replaced by the stroke color. 100% black and white remain the same.

- **Hue Shift** converts the **Key Color** (by default, the prominent color in the artwork) to the active stroke color and rotates all the other brush colors in the color wheel proportionally to that. It's like using the color wheel in the **Recolor Artwork** feature when you move all the linked colors of the artwork to a different color.

TIP **To modify an existing pattern brush, double-click the brush in the Brushes panel to open the Pattern Brush Options dialog box.**

Managing Brushes

Saving a brush lets you access it from other documents. Deleting a brush removes it from the active document and its **Brushes** panel.

Save document brushes as a library

Do the following:

1. In the **Brushes** panel, add and delete the brushes as needed.

2. Select **Save Brush Library** from the panel menu.

3. In the **Save Brushes as Library** dialog box, enter a name for the library.

4. Either accept the default path to the Illustrator **Brushes** folder (recommended) or select a specific location to save the file.

5. Click **Save**.

TIP After you save the brushes as a library, you can access the library under **User Defined**.

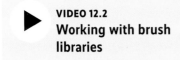

VIDEO 12.2
Working with brush libraries

Delete brushes

Do the following (**Figure 12.17**):

1. In the **Brushes** panel, select the brush you want to remove.

2. Either click the **Delete Brush** button, or select **Delete Brush** from the panel menu.

3. Click **Yes** to confirm that you want to remove the brush.

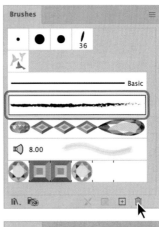

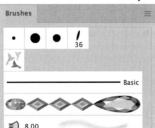

FIGURE 12.17 Deleting a brush

TIP Deleting unused brushes helps manage the file size of your document.

Painting with the Paintbrush Tool

The **Paintbrush** tool (**Figure 12.18**) uses the active fill or stroke and converts it to brush strokes as you paint with it.

Paint a path

With the **Paintbrush** tool selected, do the following (**Figure 12.19**):

1. Position the cursor where you want to begin painting.

2. Click+drag to paint the path.

3. Release the mouse.

FIGURE 12.18 The Paintbrush tool in the Essentials Classic toolbar

TIP The Brushes panel must contain brushes to use the Paintbrush tool.

FIGURE 12.19 Painting with the active color to create a stroked path

TIP Painted paths can also be adjusted by selecting and repositioning the anchor points.

TIP Edit Selected Paths must be active in the Paintbrush Tool Options dialog box in order for a path to be modified.

Modify a painted path

Do the following (**Figure 12.20**):

1. Select the painted path with the **Selection** tool.

2. With the **Paintbrush** tool, position the cursor where you want to adjust the path, and then click+drag.

3. Release the mouse.

FIGURE 12.20 Adjusting a painted path

Customize the Paintbrush tool options

Do the following:

1. Double-click the **Paintbrush** tool to open the **Paintbrush Tool Options** dialog box (**Figure 12.21**).

2. Set any of the following options:

   Adjust the **Fidelity** to adjust the curve by adding points to increase the accuracy or deleting points to increase the smoothness.

   Select **Fill New Brush Strokes** if you want to apply a fill to the path.

   Select **Keep Selected** to keep the path active after it is painted.

   Select **Edit Selected Paths** to allow the path to be modified.

   For **Within**, adjust the number of pixels to determine how close the cursor needs be to on a path in order to edit it.

3. Click **OK**.

FIGURE 12.21 The Paintbrush Tool Options dialog box

Painting with the Blob Brush Tool

The **Blob Brush** tool (**Figure 12.22**) applies calligraphic brush strokes as you paint and automatically merges overlapping strokes into a filled (usually compound) path.

> **TIP** Strokes must have the same brush fill to be merged but can have varied brush size settings.

FIGURE 12.22 Selecting the Blob Brush in the Essentials Classic toolbar

Paint a filled compound path

With the **Blob Brush** tool selected, do the following (**Figure 12.23**):

1. Set the stroke width by doing any of the following:

   In the **Control**, **Properties**, or **Stroke** panel, choose or enter a stroke weight.

   Double-click the **Blob Brush** tool to open **Blob Brush Tool Options** dialog box and adjust the brush options.

   Select a calligraphy brush from the **Brushes** panel to use its settings.

2. Position the cursor where you want to begin painting.

3. Click+drag to paint a compound path and then release the mouse.

4. (Optional) Repeat steps 1 through 3 as needed to complete the artwork.

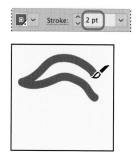

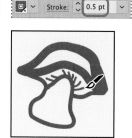

FIGURE 12.23 Painting with the Blob Brush tool using different stroke widths to create a single compound path

Customize the Blob Brush tool options

Do the following:

1. Double-click the **Blob Brush** tool to open the **Blob Brush Tool Options** dialog box (**Figure 12.24**).

2. Set any of the following options and then click **OK**:

   Select **Keep Selected** to keep the compound path active after it is painted.

   Select **Merge Only with Selections** to allow only active (usually compound) paths to be merged with new ones.

Adjust the **Fidelity** to control the precision with which the path follows your mouse or stylus movement—the smoother the setting, the simpler the path.

Adjust the **Default Brush Options** to set the calligraphic size, angle, and roundness of the brush, and whether those settings are fixed, random, or react to the stylus input on the graphic tablet.

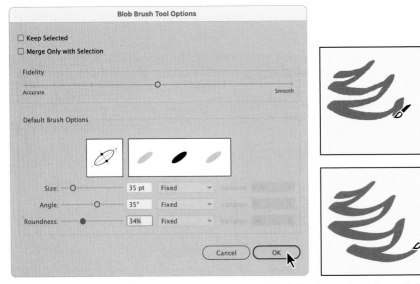

FIGURE 12.24 Customizing the Blob Brush tool options in the dialog box and painting with the applied settings

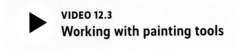

VIDEO 12.3
Working with painting tools

Sketching with the Shaper Tool

The **Shaper** tool (**Figure 12.25**) replicates traditional sketching strokes, simultaneously converting them into geometric vector shapes.

Sketch and customize shapes with the Shaper tool

With the **Shaper** tool selected, do the following (**Figure 12.26**):

1. Roughly draw a shape.

2. (Optional) Select the shape using the **Selection** tool to edit the shape or apply a fill and/or stroke.

FIGURE 12.25 The Shaper tool in the Essentials Classic toolbar

TIP The shape options available with the Shaper tool are rectangles, ellipses, diamonds, hexagons, and lines.

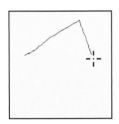
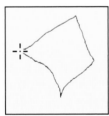
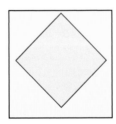
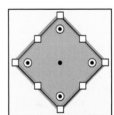

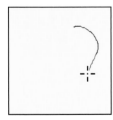
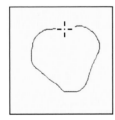
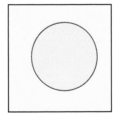
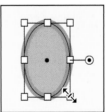

FIGURE 12.26 Sketching and customizing shapes

Sketching with the Pencil Tool

The **Pencil** tool (**Figure 12.27**) replicates traditional sketching strokes, simultaneously converting them into vector paths that are assigned the active stroke settings.

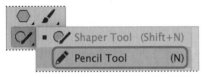

FIGURE 12.27 The Pencil tool in the Essentials Classic toolbar

Customize the Pencil tool options

Do the following:

1. Double-click the **Pencil** tool to open the **Pencil Tool Options** dialog box.

2. Set any of the following options and then click **OK**:

   Adjust the **Fidelity** to control the precision with which the path follows your mouse or stylus movement—the smoother the setting, the simpler the path.

   Select **Fill New Pencil Strokes** if you want to apply a fill to the path.

   Select **Keep Selected** to keep the path active after it is painted.

   Select and adjust **Close Paths When Ends Are Within: _ Pixels** to set how close the cursor needs to be to the start of the path to close it.

   Select **Edit Selected Paths** to allow the path to be modified.

   For **Within**, adjust the number of pixels to determine how close the cursor needs be to on a path in order to edit it.

Sketch a path with the Pencil tool

Do any or a combination of the following:

- Click+drag to create a freeform path (**Figure 12.28**).

- Press **Shift** as you click+drag to create paths constrained to 45-degree increments (**Figure 12.29**).

- Press **Alt** as you click+drag to create straight unconstrained paths.

- Click+drag and then press **Alt/Option** to close the path (**Figure 12.30**).

FIGURE 12.28 Sketching a freeform path

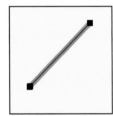

FIGURE 12.29 Pressing Shift to constrain a path

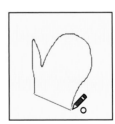

FIGURE 12.30 Pressing Alt/Option to close a path

13

Modifying Vector Objects and Paths

Modifying artwork is a user-friendly experience with the tools, panels, and commands that Illustrator provides.

In This Chapter

Modifying Objects Using Bounding Boxes

You can easily modify selected objects using the bounding box that surrounds them.

Hide or show the bounding box

Do either of the following:

- Choose **View** > **Hide Bounding Box**.
- Choose **View** > **Show Bounding Box**.

TIP The bounding box is visible by default.

Scale objects free-form

Do the following (**Figure 13.1**):

1. Hover the cursor *over* one of the bounding box handles until it displays a double-headed arrow.

2. Click+drag to resize the objects.

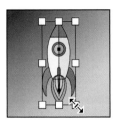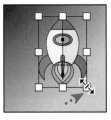

FIGURE 13.1 Click+dragging a bounding box handle point to resize objects

TIP To learn more about selecting objects, see Chapter 7.

Scale objects proportionally

Do the following (**Figure 13.2**):

1. Hover the cursor *over* one of the bounding box handles until it displays a double-headed arrow.

2. Press **Shift** as you click+drag to resize the objects.

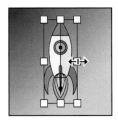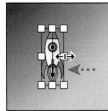

FIGURE 13.2 Pressing Shift while click+dragging a bounding box handle to proportionally resize objects

TIP Pressing **Alt/Option** as you click+drag rescales objects from the center of the bounding box.

Rotate objects free-form

Do the following (**Figure 13.3**):

1. Hover the cursor *slightly outside* one of the bounding box handles until it displays a curved double-headed arrow.

2. Click+drag to rotate the objects.

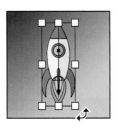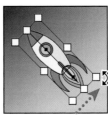

FIGURE 13.3 Click+dragging outside a bounding box anchor point to rotate objects

> **TIP** Pressing Shift as you click+drag constrains the rotation to 45-degree increments.

Reset the bounding box for rotated objects

If you need to reset the active bounding box, do the following (**Figure 13.4**):

- Choose **Object** > **Transform** > **Reset Bounding Box**.

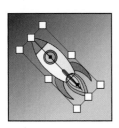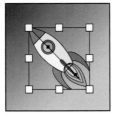

FIGURE 13.4 Rotated bounding box before and after reset

Flip objects free-form

Do the following (**Figure 13.5**):

1. Hover the cursor *over* one of the bounding box handles until it displays a double-headed arrow.

2. Click+drag past the opposite horizontal anchor point.

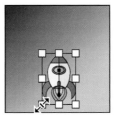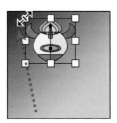

FIGURE 13.5 Click+dragging a bounding box handle to flip objects

 VIDEO 13.1
Working with bounding boxes

Modifying Objects Using Tools

The toolbar contains multiple tools for modifying objects manually or precisely.

Rotate objects manually using the Rotate tool

With the object or objects selected and the **Rotate** tool (**Figure 13.6**) active, do any of the following:

- To rotate the selection around the center point, click+drag in the direction you want to rotate (**Figure 13.7**).

- To use a different reference point for the rotation, click once to reposition the point, and then click+drag in the direction you want to rotate (**Figure 13.8**).

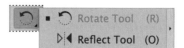

FIGURE 13.6 The Rotate tool in the Essentials Classic toolbars

TIP Pressing Shift as you click+drag constrains the rotation to 45-degree increments.

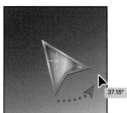

FIGURE 13.7 Rotating a selection around the center point

TIP By default, Illustrator displays the angle of rotation as you click+drag.

Rotate a copy of a selection manually using the Rotate tool

With the object or objects selected and the **Rotate** tool active, do the following (**Figure 13.9**):

- Press **Alt/Option** as you click+drag in the direction you want to rotate.

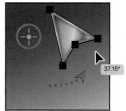

FIGURE 13.8 Repositioning a selection's reference point and then rotating it

TIP The farther from the reference point you click+drag, the greater the amount of control over the angle of rotation.

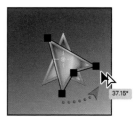

FIGURE 13.9 Creating a duplicate of a selection while rotating it

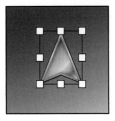

FIGURE 13.10 Rotating a selection precisely using the dialog box

TIP If your selection contains patterns, selecting Transform Patterns will rotate them with the selection. Deselecting Transform Objects rotates only the pattern within the shape.

Rotate objects precisely using the Rotate tool

With the object or objects selected, do the following (**Figure 13.10**):

1. Double-click the **Rotate** tool to open the dialog box.

2. In the **Rotate** dialog box, enter the degree of rotation.

3. Either click **OK** to rotate the selection or click **Copy** if you want the rotated selection to be a duplicate of the original.

Stop rotating selections with the Rotate tool

Do either of the following:

- Select a different tool.
- Choose **Select** > **Deselect**.

Reflect objects manually using the Reflect tool

With the object or objects selected and the **Reflect** tool active (**Figure 13.11**), do either of the following:

- Click to set the first point of the reflection axis, and then click again to set the second point (**Figure 13.12**).

- Click to set the first point of the reflection axis, and then click+drag to visually set the second point (**Figure 13.13**).

FIGURE 13.11 The Reflect tool in the Essentials Classic toolbar

TIP Pressing Shift as you click constrains the reflection to 45-degree increments.

  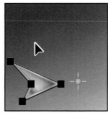 

FIGURE 13.12 Reflecting a selection by clicking to add the reflection points

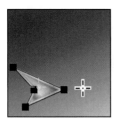 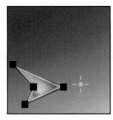 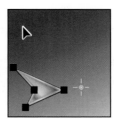 

FIGURE 13.13 Reflecting a selection by click+dragging to visually set the second reflection point

Reflect a copy of a selection manually using the Reflect tool

With the object or objects selected and the **Reflect** tool active, do the following (**Figure 13.4**):

- Press **Alt/Option** as you click or click+drag to set the reflection.

TIP By default, Illustrator displays the angle of reflection as you click+drag.

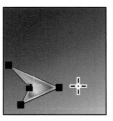 

FIGURE 13.14 Creating a duplicate of a selection while reflecting it

Reflect objects precisely using the Reflect tool

With the object or objects selected, do the following (**Figure 13.15**):

1. Double-click the **Reflect** tool to open the **Reflect** dialog box.

2. Either select **Horizontal** or **Vertical** to flip the selection along that axis, or enter a degree amount for the **Angle** of the reflection.

3. Either click **OK** to flip or reflect the selection, or click **Copy** if you want the reflected selection to be a duplicate of the original.

Stop reflecting selections with the Reflect tool

Do either of the following:

- Select a different tool.

- Choose **Select** > **Deselect**.

TIP If your selection contains patterns, choosing Transform Patterns will reflect them with the selection.

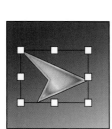

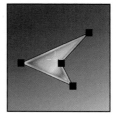

FIGURE 13.15 Reflecting a selection precisely using the dialog box

Resize objects manually using the Scale tool

With the object or objects selected and the **Scale** tool active (**Figure 13.16**), do either of the following:

- Click+drag diagonally to resize the selection based on the default center point as reference (**Figure 13.17**).

- Click once to reposition the reference point, and then click+drag diagonally.

- Press **Shift** as you click+drag diagonally to proportionally resize the selection (**Figure 13.18**).

- Press **Shift** as you click+drag *horizontally* or *vertically* to resize the selection using a single axis (**Figure 13.19**).

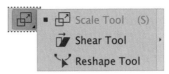

FIGURE 13.16 The Scale tool in the Essentials Classic toolbar

TIP The farther from the reference point you click+drag, the greater the amount of control over the scaling.

TIP Dragging past the reference point in either direction will flip the selection in addition to resizing it.

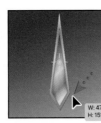

FIGURE 13.17 Manually resizing a selection

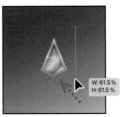

FIGURE 13.18 Resizing a selection proportionally

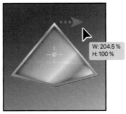

FIGURE 13.19 Resizing a selection horizontally

Resize a copy of a selection manually using the Scale tool

With the object or objects selected and the **Scale** tool active, do the following (**Figure 13.20**):

- Press **Alt/Option** as you click+drag.

FIGURE 13.20 Creating a duplicate of a selection while resizing it

TIP By default, Illustrator displays the scale percentages as you click+drag.

Resize objects precisely using the Scale tool

With the object or objects selected, do the following (**Figure 13.21**):

1. Double-click the **Scale** tool to open the dialog box.

2. In the **Scale** dialog box, either enter a percentage in the **Uniform** field to scale the selection proportionally or enter percentages in both the **Horizontal** and **Vertical** fields to resize the selection non-uniformly.

3. (Optional) Choose **Scale Corners** or **Scale Stroke & Effects** if appropriate.

4. Either click **OK** to resize the selection or click **Copy** if you want to the resized selection to be a duplicate of the original.

Set the default stroke and effects scaling option

Whether strokes and effects are scaled by default can be set in the Preferences dialog box. To set the default, do the following:

1. Choose **Edit** > **Preferences** > **General** (Windows) or **Illustrator** > **Preferences** > **General** (macOS).

2. Select or deselect **Scale Strokes & Effects**.

Stop resizing selections with the Scale tool

Do either of the following:

- Select a different tool.

- Choose **Select** > **Deselect**.

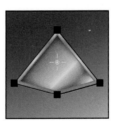

TIP If your selection contains patterns, choosing Transform Patterns will resize them with the selection.

FIGURE 13.21 Resizing a selection precisely using the dialog box

Slant or skew objects manually using the Shear tool

With the object or objects selected and the **Shear** tool active (**Figure 13.22**), do either of the following:

- To shear the selection based on the center point, click+drag in the direction you want the selection to shear (**Figure 13.23**).

- To use a different reference point for the shear, click once to reposition the point, and then click+drag in the direction you want the selection to shear (**Figure 13.24**).

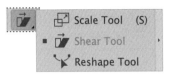

FIGURE 13.22 The Shear tool in the Essentials Classic toolbar

TIP The farther from the reference point you click+drag, the greater the control over the shearing amount.

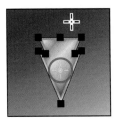 

FIGURE 13.23 Shearing a selection based on the center point

FIGURE 13.24 Repositioning a selection's reference point and then shearing it

Slant or skew a copy of a selection manually using the Shear tool

With the object or objects selected and the **Shear** tool active, do the following (**Figure 13.25**):

- Press **Alt/Option** as you click or click+drag to set the shearing.

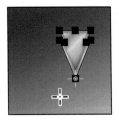 

FIGURE 13.25 Creating a duplicate of a selection while shearing it

Slant or skew objects precisely using the Shear tool

With the object or objects selected, do the following (**Figure 13.26**):

1. Double-click the **Shear** tool to open the dialog box.

2. In the **Shear** dialog box, enter a **Shear Angle** value.

3. Set the shear axis either by selecting **Horizontal** or **Vertical** or by entering an **Angle** value.

4. Either click **OK** to shear the selection or click **Copy** if you want to the sheared selection to be a duplicate of the original.

Stop shearing selections with the Shear tool

Do either of the following:

- Select a different tool.
- Choose **Select > Deselect**.

TIP If your selection contains patterns, choosing Transform Patterns will shear them with the selection.

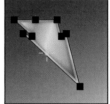

FIGURE 13.26 Shearing a selection precisely using the dialog box

Change shape contours using the Reshape tool

The **Reshape** tool (**Figure 13.27**) lets you simply change a shape's contours.

FIGURE 13.27 The Reshape tool in the Essentials Classic toolbar

TIP Smart Guides are disabled when using the Reshape tool.

Click the shape's path using the **Direct Selection** tool, Then, using the **Reshape** tool, do any of the following (**Figure 13.28**):

- Click+drag a line segment to create a new curve anchor point.

- Click+drag an anchor point to modify the contour.

- Shift-click or marquee select multiple anchor points, and then drag to modify their contours.

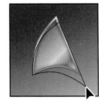

FIGURE 13.28 Selecting a shape with the Direct Selection tool and changing the contours using the Reshape tool

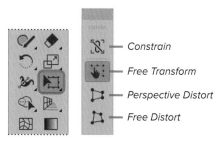

FIGURE 13.29 The Free Transform tool active in the Essentials Classic toolbar and the Free Transform widget

Constrain

Free Transform

Perspective Distort

Free Distort

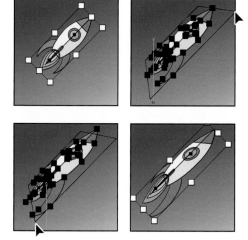

FIGURE 13.30 Applying perspective distortion

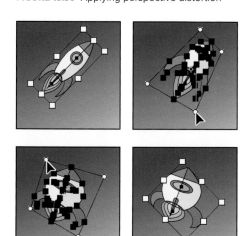

FIGURE 13.31 Applying free distortion

Distort objects using the Free Transform tool

The **Free Transform** tool lets you distort (as well as rotate, reflect, scale, and shear) selections using an independent widget (**Figure 13.29**).

With the object or objects selected and the **Free Transform** tool active, do either of the following:

- Apply perspective distortion by selecting the **Perspective Distort** button and click+dragging the corner handles (**Figure 13.30**).

- Apply free distortion by selecting the **Free Distort** button and click+dragging the corner handles (**Figure 13.31**).

TIP Distortion behavior is dependent on the bounding box orientation. If you have rotated the selection prior to distorting it, you may want to reset the bounding box (Object > Transform > Reset Bounding Box).

Modify objects using the Free Transform tool

With the object or objects selected and the **Free Transform** tool active, do any of the following:

- Rotate, reflect, scale, or shear the selection by selecting the **Free Transfrom** button and click+dragging the appropriate handles.

- Select the **Constrain** button to maintain the proportions for the free transformation.

Combine shapes using the Shape Builder tool

The **Shape Builder** tool (**Figure 13.32**) is a powerful interactive tool that lets you create complex shapes from overlapping simple ones.

With at least two overlapping shapes selected, do the following (**Figure 13.33**):

1. Select the **Shape Builder** tool.

2. Click+drag over the segments to be merged.

FIGURE 13.32 The Shape Builder tool in the Essentials Classic toolbar

 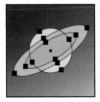 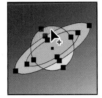 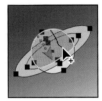

FIGURE 13.33 Click+dragging to merge overlapping shapes

Delete a segment using the Shape Builder tool

With at least two overlapping shapes selected, do the following (**Figure 13.33**):

1. Select the **Shape Builder** tool.

2. Press **Alt/Option** as you click or click+drag over the segments (**Figure 13.34**).

TIP The new shape's fill and stroke properties are dependent on which segment is selected first and whether the active properties have been updated prior to the Shape Builder tool action. They are also dependent on whether Pick Color From is set to Color Swatches or Artwork in the Shape Builder Tool Options dialog box (**Figure 13.35** on the next page).

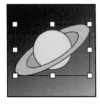  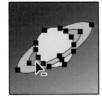 

FIGURE 13.34 Clicking to delete segments

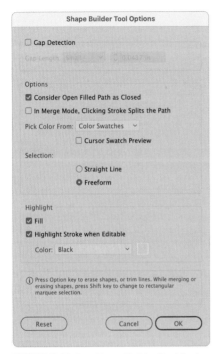

FIGURE 13.35 The Shape Builder Tool Options dialog box

Customize the Shape Builder tool behavior

Double-click the **Shape Builder** tool to open the **Shape Builder Tool Options** dialog box (**Figure 13.35**), and then do any of the following:

- Select **Gap Detection** and specify the **Gap Length** if you want Illustrator to recognize closely placed (but not over-lapping) shapes as connected.

- Select **In Merge Mode, Clicking Stroke Splits the Path** if you want the new shape to be created using the path intersection points connected by a straight line (**Figure 13.36**).

TIP Deselecting the In Merge Mode, Click-ing Stroke Splits the Path option is best for unfilled paths as it allows you to split the path at intersections without closing it.

- Choose whether the new shape color should be chosen from the artwork or the currently active color.

- Choose whether when click+dragging to select segments, the path is a free-form or straight.

- Customize the selected segment's highlight display properties.

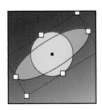 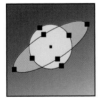 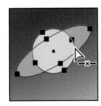 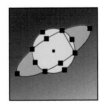 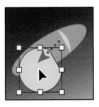

FIGURE 13.36 Clicking a path segment to create a new shape

Cutting Objects and Paths Using Tools

The cutting group of tools—the **Eraser**, **Scissors**, and **Knife** tools—lets you easily divide elements or remove unwanted segments.

FIGURE 13.37 The Eraser tools group in the Essentials Classic toolbar

Remove segments using the Eraser tool

The **Eraser** tool (**Figure 13.37**) uses free-hand strokes to remove segments from paths and shapes.

TIP Elements do not need to be selected to be erased; they only need to be unlocked.

TIP After erasing, the remaining objects become new shapes and retain their same fill and stroke properties.

With the **Eraser** tool active, do the following (**Figure 13.38**):

- Click+drag over the segments you want to erase.

FIGURE 13.38 Erasing segments from multiple elements

Customize the Eraser tool

To modify the stroke appearance, do the following (**Figure 13.39**):

1. Double-click the **Eraser** tool.

2. In the **Eraser Tool Options** dialog box, customize the calligraphic **Angle**, **Roundness**, and **Size** as needed.

3. Click **OK** to update the stroke behavior.

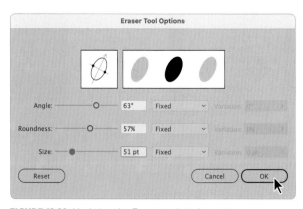

FIGURE 13.39 Updating the Eraser tool stroke options

Cut paths into segments using the Scissors tool

The **Scissors** tool (Figure 13.37) cuts paths as you click on them.

With the element selected and the **Scissors** tool active, do the following (**Figure 13.40**):

1. Click the path or an anchor point of the element to add a cut point.

2. Click the path or an anchor in a second location to separate a closed path or further split an open path.

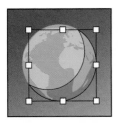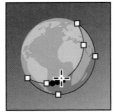
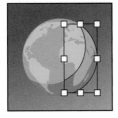

FIGURE 13.40 Cutting an object into separate segments

TIP If no element is selected, the **Scissors** tool will cut the one that is topmost.

Cut segments using the Knife tool

The **Knife** tool (Figure 13.37) cuts closed paths or filled open paths as you click+drag across them.

With the closed paths or filled open path objects selected and the **Knife** tool active, do the following (**Figure 13.41**):

- Click+drag over the element where you want to separate it.

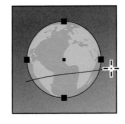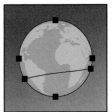
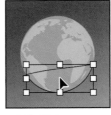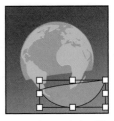

FIGURE 13.41 Cutting and separating a circle object

TIP If no element is selected, the Knife tool will cut all unlocked objects that are dragged over.

Modifying Paths Using Freeform Tools

Located under the **Shaper** tool in the Essentials Classic toolbar are a number of additional path-editing tools (**Figure 13.42**).

Adjust Bézier paths using the Smooth tool

With the object selected and the **Smooth** tool active, do the following:

- Click an anchor point where you want to smooth the path (**Figure 13.43**).
- Click+drag from the beginning to the end of the path segment you want to smooth.

Customize the Smooth tool

To modify the smoothness settings, do the following (**Figure 13.44**):

1. Double-click the **Smooth** tool.
2. In the **Smooth Tool Options** dialog box, adjust the **Fidelity** to modify the smoothness, as needed.
3. Click **OK**.

> **TIP** The greater the smoothness setting, the less the accuracy of the curve compared to the original.

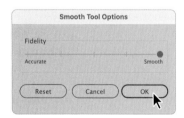

FIGURE 13.44 Adjusting the smoothness settings

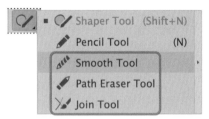

FIGURE 13.42 The path editing tools under the Shaper tool in the Essentials Classic toolbar

> **TIP** For more information about the Shaper tool, see "Sketching with the Shaper Tool" in Chapter 12.

FIGURE 13.43 Smoothing a path by clicking an anchor point

Remove path segments using the Path Eraser tool

With the object or objects selected and the **Path Eraser** tool active, do the following (**Figure 13.45**):

- Click+drag a freeform path over the segment that you want to erase.

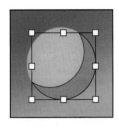
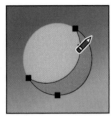
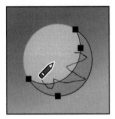
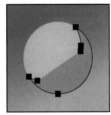

FIGURE 13.45 Erasing a path segment

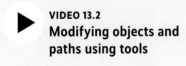

Connect open path segments and continue their curvature using the Join tool

With the path selected and the **Join** tool active, do the following (**Figure 13.46**):

- Click+drag a freeform path to connect the segments.

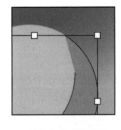
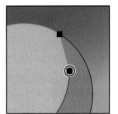
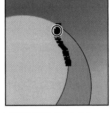

FIGURE 13.46 Joining a path segment

Connect and trim overlapping path segments using the Join tool

With the path selected and the **Join** tool active, do the following:

- Click the intersection of the overlapping paths.

Modifying Objects Using Panels

Illustrator provides numerous panels to help you precisely modify your artwork.

Transform the size and orientation of selected objects

The **Transform** panel (**Window** > **Tranform**) and **Transform** sections of the **Properties** and **Control** panels let you precisely change the size and orientation of selected objects (**Figure 13.47**).

With the object or objects selected, do any of the following:

- Change the **Reference Point** to perform the transformation (**Figure 13.48**).

- Change the position by entering new values in the **X** (horizontal) and/or **Y** (vertical) fields.

- Resize the selected objects by entering new values in the **W** (width) and/or **H** (height) fields (**Figure 13.48**).

- Lock or unlock the resize proportions by clicking the **Constrain Width and Height Proportions** toggle (**Figure 13.48**).

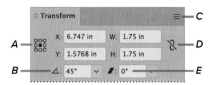

FIGURE 13.47
A. Reference Point Locator **B.** Rotate field
C. Panel menu **D.** Constrain Width and Height
Proportions toggle **E.** Shear field

- **Rotate** the selected objects by entering or selecting a new value in the field.

- **Shear** the selected objects by entering or selecting a new value in the field.

- Flip the selected objects by choosing **Flip Horizontally** or **Flip Vertically** from the **panel menu**.

- Change whether to **Scale Strokes & Effects** at the bottom of the **Transform** panel or from the panel menu.

- Change whether to **Scale Corners** at the bottom of the **Transform** panel.

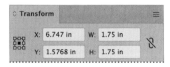

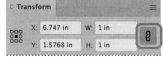

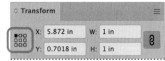

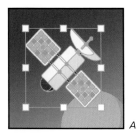

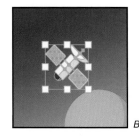

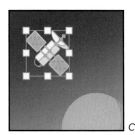

A

B

C

FIGURE 13.48
A. Original **B.** Resized with proportions constrained **C.** Resized with reference point relocated to the upper left

Modify shape properties using the Transform panel

When individual objects created using either the **Rectangle**, **Rounded Rectangle**, **Ellipse**, **Polygon**, or **Line Segment** tools are selected, the **Transform** panel displays additional modification options.

Do any of the following:

- Modify a rectangle or rounded rectangle's dimensions, rotation, corner type, and/or corner size (**Figure 13.49**).

- Modify an ellipse's dimensions, rotation, and/or pie start and end angles and/or invert settings (**Figure 13.50**).

- Modify a polygon's dimensions, rotation, number of sides, and/or corner type (**Figure 13.51**).

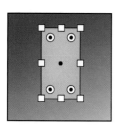
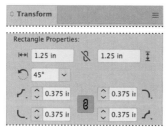
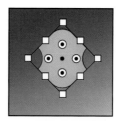

FIGURE 13.49 Modifying a rectangle's properties, including inverting some rounded corners

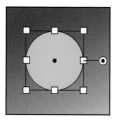
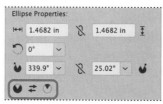
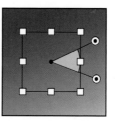

FIGURE 13.50 Modifying an ellipse's properties by adjusting and inverting (highlighted) the pie segment settings

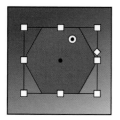

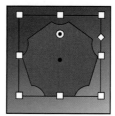

FIGURE 13.51 Modifying a polygon's properties, including decreasing the number of sides and changing the corners to inverted round

Combine and divide shape objects using the Pathfinder panel

The **Pathfinder** panel (**Window** > **Path-finder**) and **Pathfinder** sections of the **Properties** panel allow you to combine objects into new shapes or divide them into separate ones (**Figure 13.52**).

With two or more overlapping objects selected, do either of the following (**Figure 13.53**):

- Click a **Shape Modes** icon to combine the selected objects into compound paths or paths.

- Click a **Pathfinders** icon to divide or remove elements from the objects.

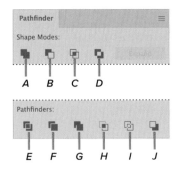

FIGURE 13.52
A. Unite
B. Minus Front
C. Intersect
D. Exclude
E. Divide
F. Trim
G. Merge
H. Crop
I. Outline
J. Minus Back

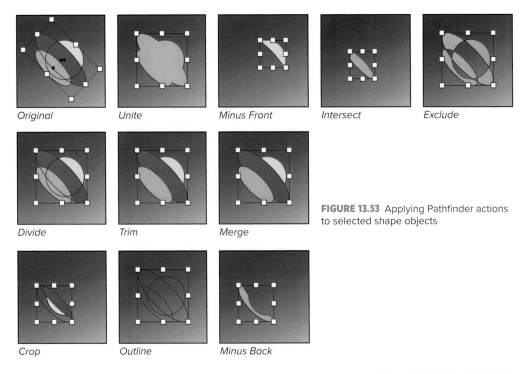

Original   Unite   Minus Front   Intersect   Exclude

Divide   Trim   Merge

FIGURE 13.53 Applying Pathfinder actions to selected shape objects

Crop   Outline   Minus Back

VIDEO 13.3
Modifying objects using panels and commands

Pathfinder panel actions defined

The number of actions in the Pathfinder panel can be overwhelming. The following are detailed descriptions of what each action does:

- **Unite** merges all the selected objects into a single one.
- **Minus Front** creates a new shape by deleting all the segments overlapping the bottom object.
- **Intersect** creates a new shape from the intersection of all involved (selected) objects.
- **Exclude** makes even-numbered overlapping elements transparent, and odd-numbered overlapping elements filled.
- **Divide** separates objects where they overlap, creating separate shapes that retain their fill appearance.
- **Trim** removes hidden segments of selected filled objects.
- **Merge** removes hidden segments of selected filled objects and merges any overlapping objects with the same fill properties.
- **Crop** converts the top selected object to the boundary for the objects under it.
- **Outline** separates the selected objects into their segments.
- **Minus Back** subtracts all shapes from the topmost one.

Convert objects into a compound shape using the Pathfinder panel

Compound shapes are containers of individual shapes that can be edited independently.

With two or more overlapping objects selected, do the following (**Figure 13.54**):

- Choose **Make Compound Shape** from the **Pathfinder** panel menu.

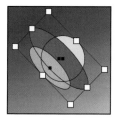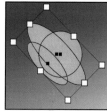

FIGURE 13.54 Converting overlapping shapes into a compound shape

Expand a compound shape using the Pathfinder panel

With the compound shape selected, do the following (**Figure 13.55**):

- Click the **Expand** button in the **Pathfinder** panel.

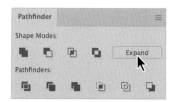

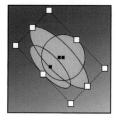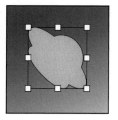

FIGURE 13.55 Expanding a compound shape

Modifying Objects Using Commands

The **Transform** commands in the **Object** menu (**Figure 13.56**) let you precisely modify objects.

Use the Object menu to modify artwork

With the objects selected, choose any of the following under **Object > Transform**:

- **Transform Again** to repeat the last transform action.
- **Move** to reposition the selection using the **Move** dialog box.
- **Rotate** to change the angle of the selection using the **Rotate** dialog box.
- **Reflect** to reflect the selection using the **Reflect** dialog box.

FIGURE 13.56 Object menu Transform commands

- **Scale** to resize the selection using the **Scale** dialog box.
- **Shear** to shear the selection using the **Shear** dialog box.
- **Transform Each** to modify selected objects and groups individually using the **Transform Each** dialog box (**Figure 13.57**).

TIP For more information about the Rotate, Reflect, Scale, and Shear dialog boxes, see "**Modifying Objects Using Tools**" in this chapter.

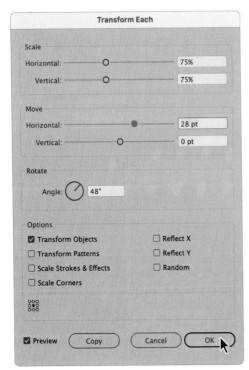

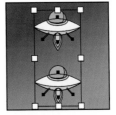

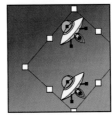

FIGURE 13.57 Transforming groups of objects individually

Transforming Objects

Illustrator transformation tools and options provide endless possibilities to transform simple elements into visually rich artwork.

In This Chapter

Using Liquify Tools to Reshape Objects

The liquify group of tools (**Figure 14.1**) provides preset distortion options, allowing you to easily add distortions, such as wrinkles and twirls, to elements.

TIP Selecting objects isolates them for the liquify actions. If no object is selected, liquify actions will be applied to all unlocked elements within the cursor boundary.

Use the Warp tool to mold object paths

With the **Warp** tool active, do the following (**Figure 14.2**):

- Click+drag over the portion of the artwork in the direction you want to warp.

Customize the Warp tool

Double-click the **Warp** tool to open the **Warp Tool Options** dialog box, and then do any of the following (**Figure 14.3**):

- Modify the size of the cursor by changing the **Width** and **Height** settings.
- Modify the orientation of an asymmetric cursor by changing the **Angle**.
- Specify how quickly the warp is applied by adjusting the **Intensity**. The lower the value, the slower the change occurs.
- Specify the spacing between the added warp action points by adjusting the **Detail**. The lower the value, the greater the distance.
- Specify how much to reduce the amount of unneeded action points by adjusting the **Simplify** values.

FIGURE 14.1 Liquify tools located under the Width tool in the Essentials Classic toolbar

FIGURE 14.2 Warping a star shape

FIGURE 14.3 The Warp Tool Options dialog box

FIGURE 14.4 Swirling a star shape

```
                Twirl Tool Options

Global Brush Dimensions

    Width: ⌃ 60 pt          ⌄

    Height: ⌃ 60 pt         ⌄

    Angle: ⌃ 0°             ⌄

   Intensity: ⌃ 50%         ⌄

              ☐ Use Pressure Pen


Twirl Options

 Twirl Rate: ————————○———————— 40°

 ☑ Detail:   ——○————————————— 2

 ☑ Simplify: ————————○———————— 50


 ☑ Show Brush Size

 ⓘ  Brush Size may be interactively changed by holding down the
    Option Key before clicking with the tool.

 ( Reset )        ( Cancel )    ( OK )
```

FIGURE 14.5 The Twirl Tool Options dialog box

Use the Twirl tool to add swirls to objects

With the **Twirl** tool active, do the following (**Figure 14.4**):

- Click+drag over the portion of the art-work you to which want to add a swirl.

TIP The slower you drag over the artwork, the more pronounced the tool applies the twirl.

Customize the Twirl tool

Double-click the **Twirl** tool to open the **Twirl Tool Options** dialog box, and then do any of the following (**Figure 14.5**):

- Modify the size of the cursor by chang-ing the **Width** and **Height** settings.

- Modify the orientation of an asymmetric cursor by changing the **Angle**.

- Specify how quickly the twirl is applied by adjusting the **Intensity**. The lower the value, the slower the change occurs.

- Determine rotation and rate of the twirl by adjusting the **Twirl Rate**. Positive values twirl clockwise and negative values counterclockwise. The closer to zero, the slower the twirl is applied.

- Specify the spacing between twirl action points by adjusting the **Detail**. The lower the value, the greater the distance.

- Specify how much to reduce the amount of unneeded action points by adjusting the **Simplify** values.

Use the Pucker tool to distort an object's appearance

With the **Pucker** tool active, do the following (**Figure 14.6**):

- Click or click+drag over the portion of the artwork you want to distort.

Customize the Pucker tool

Double-click the **Pucker** tool to open the **Pucker Tool Options** dialog box, and then do any of the following (**Figure 14.7**):

- Modify the size of the cursor by changing the **Width** and **Height** settings.

- Modify the orientation of an asymmetric cursor by changing the **Angle**.

- Specify how quickly the pucker is applied when dragging by adjusting the **Intensity**. The lower the value, the slower the change occurs.

- Specify the spacing between pucker action points by adjusting the **Detail**. The lower the value, the greater the distance.

- Specify how much to reduce the amount of unneeded action points by adjusting the **Simplify** values.

FIGURE 14.6 Clicking a star shape segment to deflate it

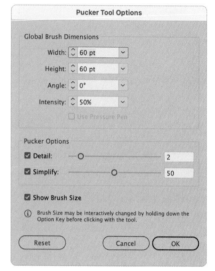

FIGURE 14.7 The Pucker Tool Options dialog box

FIGURE 14.8 Clicking a star shape segment to inflate it

FIGURE 14.9 The Bloat Tool Options dialog box

Use the Bloat tool to distort an object's appearance

With the **Bloat** tool active, do the following (**Figure 14.8**):

- Click or click+drag over the portion of the artwork you want to distort.

Customize the Bloat tool

Double-click the **Bloat** tool to open the **Bloat Tool Options** dialog box, and then do any of the following (**Figure 14.9**):

- Modify the size of the cursor by changing the **Width** and **Height** settings.

- Modify the orientation of the cursor by changing the **Angle**.

- Specify how quickly the bloat is applied when dragging by adjusting the **Intensity**. The lower the value, the slower the change occurs.

- Specify the spacing between bloat action points by adjusting the **Detail**. The lower the value, the greater the distance.

- Specify how much to reduce the amount of unneeded action points by adjusting **Simplify** values.

Use the Scallop tool to add random curved details to an object

With the **Scallop** tool active, do the following (**Figure 14.10**):

- Click or click+drag over the portion of the artwork in the direction you want to add random curves.

Customize the Scallop tool

Double-click the **Scallop** tool to open the **Scallop Tool Options** dialog box, and then do any of the following (**Figure 14.11**):

- Modify the size of the cursor by changing the **Width** and **Height** settings.

- Modify the orientation of an asymmetric cursor by changing the **Angle**.

- Specify how quickly the scallop is applied when dragging by adjusting the **Intensity**. The lower the value, the slower the change occurs.

- Modify how near the scallops are placed in relation to the object's outline by adjusting the **Complexity**.

- Specify the spacing between scallop action points by adjusting the **Detail**. The lower the value, the greater the distance.

- Enable or disable **Brush Affects Anchor Points**, **Brush Affects In Tangent Handles**, or **Brush Affects Out Tangent Handles** to allow the tool to change these properties.

FIGURE 14.10 Click+dragging with the Scallop tool around an object to add random curved details to the edges

FIGURE 14.11 The Scallop Tool Options dialog box

FIGURE 14.12 Click+dragging with the Crystallize tool around an object to add random spiked details to the edges

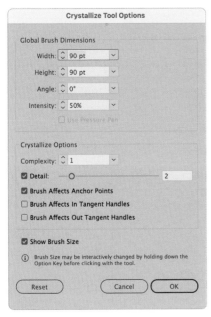

FIGURE 14.13 The Crystallize Tool Options dialog box

Use the Crystallize tool to add random spiked details to an object

With the **Crystallize** tool active, do the following (**Figure 14.12**):

- Click or click+drag over the portion of the artwork in the direction you want to add random spikes.

Customize the Crystallize tool

Double-click the **Crystallize** tool to open the **Crystallize Tool Options** dialog box, and then do any of the following (**Figure 14.13**):

- Modify the size of the cursor by changing the **Width** and **Height** settings.

- Modify the orientation of an asymmetric cursor by changing the **Angle**.

- Specify how quickly the crystallization is applied when dragging by adjusting the **Intensity**. The lower the value, the slower the change occurs.

- Modify how near the spikes are placed in relation to the object's outline by adjusting the **Complexity**.

- Specify the spacing between crystallize action points by adjusting the **Detail**. The lower the value, the greater the distance.

- Enable or disable **Brush Affects Anchor Points**, **Brush Affects In Tangent Handles**, or **Brush Affects Out Tangent Handles** to allow the tool to change these properties.

Use the Wrinkle tool to add random wavy details to an object

With the **Wrinkle** tool active, do the following (**Figure 14.14**):

- Click or click+drag over the portion of the artwork in the direction you want to add random waves.

Customize the Wrinkle tool

Double-click the **Wrinkle** tool to open the **Wrinkle Tool Options** dialog box, and then do any of the following (**Figure 14.15**):

- Modify the size of the cursor by changing the **Width** and **Height** settings.

- Modify the orientation of an asymmetric cursor by changing the **Angle**.

- Specify how quickly the wrinkle is applied when dragging by adjusting the **Intensity**. The lower the value, the slower the change occurs.

- Modify the distance percentage for how far the **Horizontal** and **Vertical** control point distances are.

- Modify how near the action results are in relation to the object's outline by adjusting the **Complexity**.

- Specify the spacing between wrinkle action points by adjusting the **Detail**. The lower the value, the greater the distance.

- Enable or disable **Brush Affects Anchor Points**, **Brush Affects In Tangent Handles**, or **Brush Affects Out Tangent Handles** to allow the tool to change these properties.

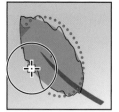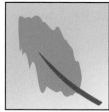

FIGURE 14.14 Click+dragging with the Wrinkle tool around an object to add random wavy details to the edges

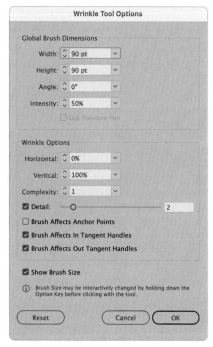

FIGURE 14.15 The Wrinkle Tool Options dialog box

FIGURE 14.16 The Blend tool in the Essentials Classic toolbar

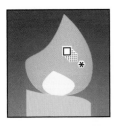 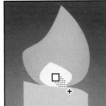

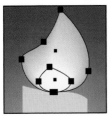 

FIGURE 14.17 Clicking objects to create a blended fill appearance

TIP Selecting objects before applying Blend tool actions makes it easier to see their anchor points.

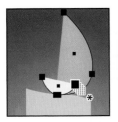 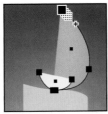

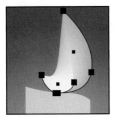 

FIGURE 14.18 Clicking object anchor points for two open path objects to create a blended fill appearance

Blending Objects

The **Blend** tool (**Figure 14.16**) and menu commands let you use two objects to create smooth color fills between them, evenly create and distribute new shapes between the two objects.

Use the Blend tool to create a blended fill visual between two objects

With the **Blend** tool active, do either of the following:

- Click each object (not their anchor points) to sequentially blend them with no rotation (**Figure 14.17**).
- Click an anchor point for each object to blend them using those points as reference (**Figure 14.18**).

TIP The Blend tool cursor changes from a white square to a black square when it is over an anchor point.

Use the Blend command to create a blended fill visual between two objects

With the two objects selected, do the following:

- Choose **Object** > **Blend** > **Make**.

Release blended objects

With the blended objects selected, do the following:

- Choose **Object** > **Blend** > **Release**.

Expand blended objects

With the blended objects selected, do the following:

- Choose **Object** > **Blend** > **Expand**.

Access the Blend Options dialog box

Do any of the following:

- Double-click the **Blend** tool.
- Choose **Object > Blend > Blend Options**.
- With blended objects selected, click the **Blend Options** button under **Quick Actions** in the **Properties** panel (**Figure 14.19**).

Customize Blend options

In the **Blend Options** dialog box, do any of the following:

- Under **Spacing**, choose **Smooth Color** to let Illustrator automatically calculate and apply the number of steps for a smoothly blended fill or stroke.
- Under **Spacing**, choose **Specified Steps** and enter an appropriate number to determine the number of blended objects between the two original ones (**Figure 14.20**).
- Under **Spacing**, choose **Specified Distance** and enter an appropriate number to specify the distance from the edge of one object and the corresponding edge of the next.
- Under **Orientation**, choose **Align to Page** to set the blending action perpendicular to the page's x-axis.
- Under **Orientation**, choose **Align to Path** to set the blending action perpendicular to the blend's spine.

TIP The path that the blended objects are aligned along is called the spine.

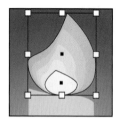

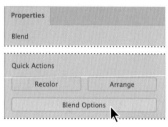

FIGURE 14.19 Accessing the Blend Options dialog box from the Properties panel with blended objects selected

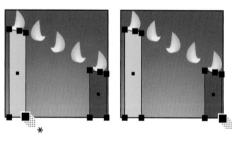

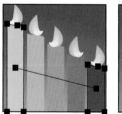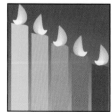

FIGURE 14.20 Creating distributed shapes using the Specified Steps option

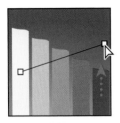

FIGURE 14.21 Adjusting a spine anchor point

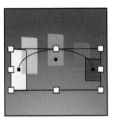 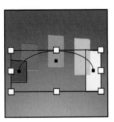

FIGURE 14.22 Replacing a spine

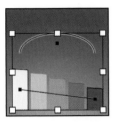 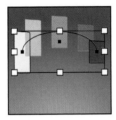

FIGURE 14.23 Reversing the blend along the spine

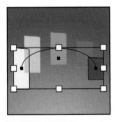 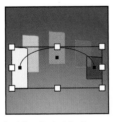

FIGURE 14.24 Reversing the blended object stacking order

Adjust blended objects spine

With the blended objects *deselected*, do the following (**Figure 14.21**):

1. Using the **Direct Selection** tool, click to select a spine anchor point.

2. Drag the spine anchor point to the needed location.

Replace the spine with a different path

Do the following (**Figure 14.22**):

1. Select the object to use for the new spine and the blended objects.

2. Choose **Object** > **Blend** > **Replace Spine**.

TIP Spines should always be open paths to prevent unexpected results.

Reverse the blend along the spine

With the blended objects selected, do the following (**Figure 14.23**):

- Choose **Object** > **Blend** > **Reverse Spine**.

Reverse the blended object stacking order

With the blended objects selected, do the following (**Figure 14.24**):

- Choose **Object** > **Blend** > **Reverse Front To Back**.

 VIDEO 14.1
Working with liquify tools and blending objects

Masking Artwork

In Illustrator, *clipping masks* are vector shapes that hide the artwork beyond their boundaries. The object that is used for the mask is called the *clipping path*. Once created, the clipping mask and related artwork it hides are called a *clipping set*.

Create a clipping mask to hide parts of layers or groups

When you want to hide multiple objects, it is best to use the **Layers** panel.

Do the following (**Figure 14.26**):

1. In the **Layers** panel, organize the objects to mask within the same layer.

2. Create the vector object to use at the clipping path and place it at the top of the layer.

3. With the layer highlighted in the **Layers** panel, click the **Make/Release Clipping Mask** button at the bottom of the panel.

> **TIP** You can also use the Layers panel to mask individual objects.

Create a clipping mask to hide part of an object using commands

To mask individual objects, do the following (**Figure 14.25**):

1. Create the vector object to use as the clipping path and position it over the object you want to mask.

2. Select the object to be masked and the clipping path object.

3. Choose **Object** > **Clipping Mask** > **Make**.

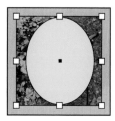

FIGURE 14.25 Creating a clipping mask for an image object

> **TIP** To learn more about placing image files in Illustrator, see Chapter 17, "Importing Assets."

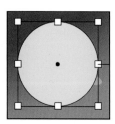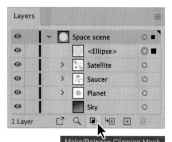

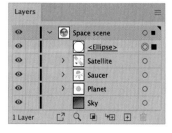

FIGURE 14.26 Creating a clipping mask for multiple objects using the Layers panel

FIGURE 14.27 Adding a stroke to a mask and modifying its shape

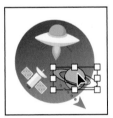
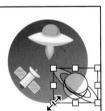

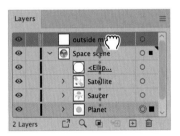
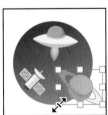

FIGURE 14.28 Selecting and modifying a clipping set object

Modify a clipping mask using commands

With the clipping set selected, do the following:

1. Choose **Object** > **Clipping Mask** > **Edit Mask**.

2. Apply a stroke or fill, or use the **Direct Selection** tool to modify the mask (**Figure 14.27**).

TIP You can also modify the mask's fill or stroke by selecting it using the Direct Selection tool.

Modify a clipping set using the Layers panel

In the **Layers** panel, do any of the following:

- Select an object or path, and use the **Selection** or **Direction Selection** tools to modify it (**Figure 14.28**).

- Add or remove objects by dragging them into or out of the layer (**Figure 14.29**).

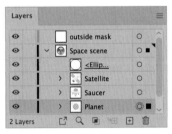

FIGURE 14.29 Removing an object from a clipping set and placing it in the layer above it

Applying Transparency and Blending Modes

Transparency and opacity modifications can be applied using a number of options.

Modify opacity settings

With the objects, layers, or groups selected, do the following (**Figure 14.30**):

- In the **Opacity** section of the **Properties**, **Control**, **Appearance**, or **Transparency** panel, enter a new value or adjust the value using the slider.

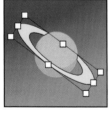
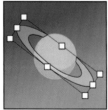

FIGURE 14.30 Reducing a selected object's opacity

Access the Transparency panel

The **Transparency** panel (**Figure 14.31**) provides several options for setting opacity, applying blending modes, and creating opacity masks.

FIGURE 14.31
A. Blending Mode menu
B. Opacity Mask section

To access the panel, do either of the following:

- Click the word **Opacity** in the **Properties**, **Control**, or **Appearance** panel.
- Choose **Window** > **Transparency**.
- In the **Essentials Classic** workspace, click the **Transparency** panel thumbnail (**Figure 14.32**).

FIGURE 14.32 Clicking the Transparency panel thumbnail

Apply a blending mode to objects

With the objects for blending selected, do the following:

- In the **Transparency** panel, choose an option from the **Blending Mode** menu (**A** in Figure 14.31).

About blending modes

Blending modes (**Figure 14.33**) determine how the overlapping colors interact with each other and any underlying elements.

TIP The object's color mode (RGB, CMYK, spot) affects how blending modes behave, and some may not work at all with spot colors.

The colors used to blend objects are **blend color**, which is the selected object's original color; **base color**, which is the color below the blend objects; and **resulting color**, which is the end blended color.

Normal is the default mode, in which there is no interaction between the blend and base colors.

Darken uses the darkest color from either the blend or base color and replaces any colors that are lighter. Areas darker than the blend color do not change.

Multiply multiplies the blend and base colors, always resulting in a darker color.

Color Burn reflects the blend color by darkening the base color.

Lighten uses the lightest color from either the blend or base color to replace any colors that are darker. Areas lighter than the blend color do not change.

Screen multiplies the blend and base inverse colors, always resulting in a lighter color.

Color Dodge reflects the blend color by lightening the base color.

Overlay, depending on the base color, either screens or multiplies the colors to reflect the dark and light variants in the original color.

Soft Light, depending on the base color, either darkens or lightens the colors, similar to a diffused spotlight.

Hard Light, depending on the base color, either screens or multiplies the colors, similar to a harsh spotlight.

Difference subtracts the greater value color, either the blend or the base, from the other.

Exclusion is similar to Difference but with lower contrast results.

Hue combines the hue of the blend color with the saturation and luminosity of the base color.

Saturation combines the luminance and hue of the base color and the saturation of the blend color.

Color combines the luminosity of the base color with the hue and saturation of the blend color.

Luminosity combines the hue and saturation of the base color with the luminosity of the blend color.

FIGURE 14.33 Illustrator blending modes applied to CMYK objects

Create an opacity mask

Opacity masks use images or grayscale objects to apply masks with varying degrees of transparency. Black is transparent, and white is opaque.

TIP If you want to use multiple objects as a mask, group them.

Do the following:

1. Place the object or image to be used for the mask on top of the artwork you want to include (**Figure 14.34**).

2. Select all the elements to be included, including the opacity masking object.

3. In the **Transparency** panel, click **Make Mask** (**Figure 14.35**).

4. (Optional) Deselect **Clip** to show the artwork outside the mask boundaries and/or select **Invert Mask** to reverse the mask values (**Figure 14.36**):

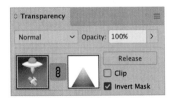

FIGURE 14.36 Deselecting Clip and then selecting Invert Mask

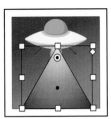

FIGURE 14.34 Placing a gradient-filled object over artwork to be masked

FIGURE 14.35 Creating an opacity mask for selected objects

Undo an opacity mask

Do the following:

- With opacity mask group selected, click **Release** in the **Transparency** panel.

TIP Applying an opacity mask automatically groups all the objects.

 VIDEO 14.2
Working with masks and transparency

Applying Gradients

The **Gradient** tool and panel options (**Figure 14.37**) let you add gradual blends between colors to fills and stokes. You can choose gradient presets provided with Illustrator or create one of your own.

> **TIP** If a document's gradient panel has been modified, the provided preset gradient swatches may have been removed. However, you can always access them from a new blank document or using the libraries.

Apply a preset gradient to an object

With the object selected, do the following (**Figure 14.38**):

1. Make sure that either the fill or stroke color box is active, depending on which one you want to apply the gradient to.

2. In the **Swatches** panel (**Window > Swatches**), click the gradient swatch.

 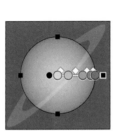 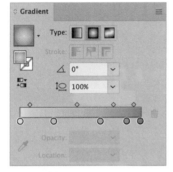

> **TIP** The Gradient panel is also accessible in the Control and Properties panels when the selected object contains a gradient or when the gradient tool is active.

FIGURE 14.37 Applying a radial gradient fill to an object

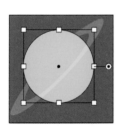 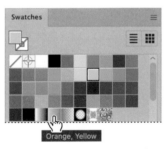 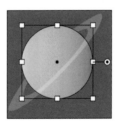

Orange, Yellow

FIGURE 14.38 Applying a gradient to an object using the Swatches panel

Change the type of gradient applied

Under **Type** in the **Gradient** panel, choose any of the following (**Figure 14.39**):

- **Linear** blends the colors in a straight line.
- **Radial** blends the colors in a circular pattern.
- **Freeform** blends the colors within the object using paths and points positioned freely inside the object.

Activate the Gradient Annotator

With the gradient object selected, do either of the following:

- Click the **Gradient** tool.
- In the **Gradient** panel, click **Edit Gradient** (**Figure 14.40**).

TIP The Gradient Annotator does not appear with gradient strokes.

TIP Clicking the Edit Gradient button in the Gradient panel automatically activates the Gradient tool.

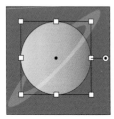
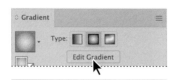
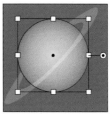

FIGURE 14.39 Changing an object's gradient type from Linear to Radial

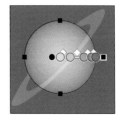

TIP In editing mode, a Gradient Annotator slider is visible for linear and radial gradients and mirrors the midpoints and color stops in the Gradient panel.

FIGURE 14.40 Activating editing options for a radial gradient

TIP The gradient annotator can be hidden or shown by choosing View > [Show/Hide] Gradient Annotator.

Modify a linear or radial gradient position and angle using the Gradient tool

With the gradient object selected and the **Gradient** tool active, do the following (**Figure 14.41**):

- Click+drag to set the new beginning and ending points.

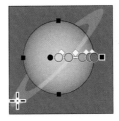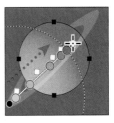

FIGURE 14.41 Repositioning a gradient using the Gradient tool

Modify a linear or radial gradient using the Gradient panel

In the **Gradient** panel (**Figure 14.42**), do any of the following (**Figure 14:43**):

- Choose a different gradient from the **Active Gradient** menu.

- Reverse the direction of the gradient by clicking the **Reverse Gradient** button.

- Click+drag to reposition the **Midpoint** icons.

- Click+drag to reposition the **Color Stop** icons.

- Add a color stop by clicking a position along the bottom of the Gradient Annotation bar.

- Delete a color stop by click+dragging it away from the **Gradient Annotator** bar, pressing **Delete**, or clicking the **Delete Stop** button.

- Change the active color of a color stop by either clicking the **Color Picker** button to choose a color or double-clicking the color stop.

- Adjust the **Opacity** of a color stop.

- Adjust the **Location** of a color stop.

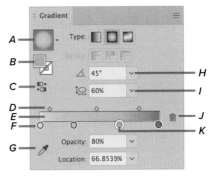

FIGURE 14.42
A. Active gradient **B.** Selected Stop color
C. Reverse Gradient button **D.** Midpoint icon
E. Gradient Annotator bar **F.** Color Stop icon
G. Color Picker button **H.** Angle field
I. Aspect Ratio field **J.** Delete Stop button
K. Active Color Stop icon

TIP The Gradient Annotator does not need to be active to make adjustments to selected gradient objects using the Gradient panel.

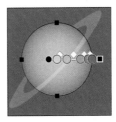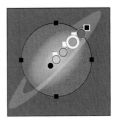

FIGURE 14.43 Before and after applying the gradient edits in Figure 14.40

Customize a freeform gradient using points

With a freeform gradient applied to the object and **Points** active in the **Gradient** panel, do any of the following:

- Modify a color stop using the **Gradient** panel (**Figure 14.44**).

- Add a color stop by clicking inside the object.

- Delete a color stop by selecting it and pressing **Delete** or clicking the **Delete Stop** button.

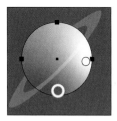

FIGURE 14.44 Adjusting a freeform gradient color stop point

Customize a freeform gradient using lines

With a freeform gradient applied to the object and **Lines** active in the **Gradient** panel, do any of the following:

- Add connected color stops by selecting a color stop and then clicking inside the object (**Figure 14.45**).

- Modify a color stop color and options settings using the **Gradient** panel (**Figure 14.46**).

- Delete a color stop by selecting it and pressing **Delete** or clicking the **Delete Stop** button.

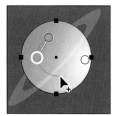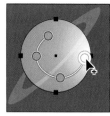

FIGURE 14.45 Connecting a freeform gradient using lines

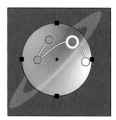

FIGURE 14.46 Modifying a freeform gradient line point

TIP Freeform gradient lines cannot overlap themselves.

Apply a new gradient

With the object selected, do the following (**Figure 14.47**):

1. Select the **Gradient** tool.

2. In either the **Gradient** panel or the **Gradient** section of the **Control** or **Properties** panel, click the gradient **Type** you want to apply.

3. (Optional) Modify the gradient, as needed.

Save a linear or radial gradient

With the gradient object selected, do the following (**Figure 14.48**):

1. In the **Swatches** panel, either click the **New Swatch** button or select **New Swatch** from the panel menu.

2. In the **New Swatch** dialog box, enter a name for the swatch, and then click **OK**.

TIP Linear and radial gradients can also be saved to the Swatches panel using the Gradient panel menu.

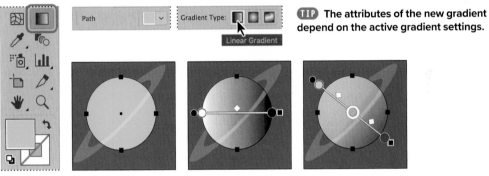

TIP The attributes of the new gradient depend on the active gradient settings.

FIGURE 14.47 Applying a new gradient to an object

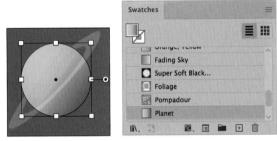

TIP Freeform gradients can be saved as graphic styles. To learn more, see Chapter 15, "Adding Visual Effects."

FIGURE 14.48 A new gradient saved as a swatch

TIP To learn more about saving swatches, see "Using the Swatches Panel" in Chapter 4, "Working with Color."

Apply gradients to strokes

With the object selected, do the following (**Figure 14.49**):

1. Make sure the stroke color box is active.

2. Choose a gradient from the **Gradient** or **Swatches** panel.

3. Select either **Linear** or **Radial** for the gradient **Type**.

4. Select a style next to the word **Stroke** (**Figure 14.50**).

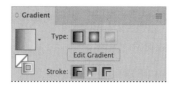

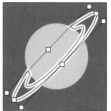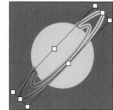

FIGURE 14.49 Applying a gradient to a stroke

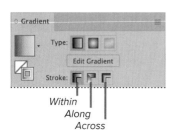

Within
Along
Across

Linear Within   *Linear Along*   *Linear Across*

Radial Within   *Radial Along*   *Radial Across*

FIGURE 14.50 Gradient stroke styles

> ▶ **VIDEO 14.3**
> **Working with gradients**

FIGURE 14.51 Live Paint tools located under the Shape Builder tool in the Essentials Classic toolbar

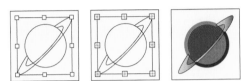

FIGURE 14.52 Creating a Live Paint Group from selected objects and attributes applied to the edges and faces

Expanding a Live Paint group

Expanding a Live Paint group flattens the faces and edges while retaining the visual similarities of the Live Paint group.

With the group selected, do the following:

- Choose **Object** > **Live Paint** > **Expand**.

Release a Live Paint group

Releasing removes the Live Paint group and reverts to paths with no fill and a half-point black stroke.

With the group selected, do the following:

- Choose **Object** > **Live Paint** > **Release**.

Working with Live Paint

Live Paint tools (**Figure 14.51**) and commands let you create *Live Paint groups* from vector artwork to which you can add color, gradients, and patterns. *Live Paint groups* retain most of the vector drawing and editing capabilities, but treat all the paths as if they are on the same flat level of the surface.

Rather than *strokes* and *fills*, the paintable portions of Live paint groups are called *edges* and *faces*.

Create a Live Paint group

With the vector objects selected, do either of the following (**Figure 14.52**):

- Select the **Live Paint Bucket** tool and click the selection.
- Choose **Object** > **Live Paint** > **Make**.

Prepare objects for Live Paint conversion

Some objects require additional actions prior to converting them. Do any of the following:

- For objects that did not convert directly, choose **Object** > **Expand**.
- For type objects, choose **Type** > **Create Outlines**.
- For raster elements, choose **Object** > **Live Trace** > **Make and Convert to Live Paint**.

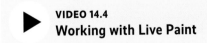

VIDEO 14.4
Working with Live Paint

Use the Live Paint Bucket tool to apply attributes to a Live Paint edge or face

The **Live Paint Bucket** adds the current fill attribute to faces within Live Paint groups or the current stroke properties to edges.

With the fill or stroke attributes you want to apply chosen, do any of the following:

- Click an edge or face (**Figure 14.53**).

- Click+drag across multiple edges or faces to paint more than one at once.

- Double-click an edge or face to apply the attribute to all unassigned adjacent elements.

- Triple-click an edge or face to stroke or fill all elements with the same attributes.

FIGURE 14.53 Using the Live Paint Bucket tool to apply a fill to a face element

TIP If the Swatches panel is used to select the fill (and Cursor Swatch Preview is selected in the Live Paint Bucket Options dialog box), the cursor displays three swatches, with the active swatch in the middle. To toggle through the swatches, press the Left and/or Right Arrow keys.

Customize the Live Paint Bucket or Live Paint Selection Tool options

Do the following:

- Double-click the tool to open its options dialog box.

TIP To apply fill to edges with the Live Paint Bucket tool, select the Paint Stokes in the Live Paint Bucket Options dialog box.

Use the Live Paint Selection tool to apply attributes to a Live Paint face or edge

The **Live Paint Selection** tool lets you choose edges and faces within Live Paint groups.

Do any of the following, and then change the stroke or fill settings (**Figure 14.54**):

- Click a face or edge.

- Click+drag a marquee around the items you want to select.

- Double-click a face to select all contiguous faces that are not separated by a painted edge.

- Triple-click a face or edge to select items with the same fill or stroke.

- Press **Shift**-click or **Shift**-click+drag a marquee to add or remove items from the selection.

TIP Select > **Same commands also work with the Live Paint Selection tool.**

Clicking to select an edge and applying a stroke

Clicking to select a face and applying a gradient fill

Using a marquee selection to apply a fill and stroke

FIGURE 14.54 Using the Live Paint Selection tool to select and apply attributes

Adding Visual Effects

Effects let you add visually interesting appearances to your artwork and easily manage them.

Applying Illustrator Effects

Effects are located under the **Effect** menu or the **fx** menu button in either the **Properties** or **Appearance** panel when objects are selected (**Figure 15.1**).

Effects are organized into two categories: **Illustrator Effects** are primarily *vector* based, and all **Photoshop Effects** are *raster* based.

FIGURE 15.1
The Effects menu options in the Properties panel

Applying effects vs. transforming objects

Effects applied using the **Effect** command menus appear as attributes that can be managed from the **Properties** or **Appearance** panels.

Several of the vector **Effect** menu groups (**Convert to Shape**, **Distort & Transform**, **Path**, **Pathfinder**, and **Warp**) mirror transformation tools and the commands found under the **Object** menu. The difference is that applying transformations using tools or the **Object** menu alters the actual object, whereas using the **Effect** menu commands alters only the appearance, leaving the original object intact.

Apply a 3D (Classic) effect to objects

With the objects selected, do the following:

1. Choose **Effect** > **3D and Materials** > **3D (Classic)** > *[effect name]*, or click the **fx** menu button in either the **Properties** or **Appearance** panel and choose **3D (Classic)** > *[effect name]*.

2. In the dialog box, customize the settings as needed, and then click **OK**.

> **TIP** If you want multiple objects to occupy the same 3D space, group them (**Object** > **Group**) before applying the 3D effect.

FIGURE 15.2 Opening the 3D and Effects panel by applying the Inflate effect to a selected group of objects

TIP As of the Illustrator 2022 release, 3D and Materials was a Technology Preview feature. This means that while included with the release, it is still considered in development with limited capabilities and documentation. For more information, click the Technology Preview link at the bottom of the panel.

Access the 3D and Materials panel by applying a 3D effect

The **3D and Materials** panel is a Technology Preview feature that allows you to easily create rich three-dimensional effects, including surface materials and lighting.

With the objects selected, do either of the following (**Figure 15.2**):

- Choose **Effect** > **3D and Materials** > *[effect name]*.

- Click the **fx** menu button in either the **Properties** or **Appearance** panel and choose **3D and Materials** > *[effect name]*.

Modify an object's 3D Type

In the **Object** section of the **3D and Materials** panel, choose or adjust any of the following:

- **Plane** orients a flat object in 3D space, similar to holding and rotating a piece of paper.

- **Extrude** adds depth to the object by extending it along its z axis.

- **Revolve** uses a circular direction to sweep the object around.

- **Inflate** adds depth by puffing up the object.

- **Depth** determines the object's z axis value.

- **Cap** determines if the object's appearance is hollow or solid.

- **Bevel** determines the type of edge applied to the end of the z axis of the object.

- **Rotation Presets** apply precalculated vertical, horizontal, and circular values for the axis, direction, and isometrics.

- **Volume** determines the strength of the inflation.

Modify a 3D object's surface material appearance

In the **Materials** section of the **3D and Materials** panel, do any of the following:

- Under **All Materials**, choose an appearance from either the **Base Materials** (default) or **Adobe Substance Materials**, which contains preset customizable surfaces provided with Illustrator (**Figure 15.3**).

- Under **Material Properties**, adjust the appearance values for the object's applied surface.

TIP The Default material option retains the object's colors and adds glossiness.

FIGURE 15.3 Applying an Adobe Substance Material to a 3D object's surface

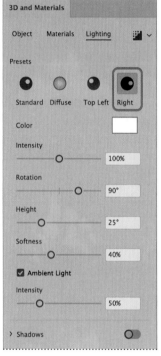

FIGURE 15.4 Applying a lighting preset to a 3D object

Modify the lighting for a 3D object

In the **Lighting** section of the **3D and Materials** panel, do any of the following:

- Choose a lighting **Preset** to quickly apply lighting effects preconfigured for direction and intensity (**Figure 15.4**).

- Change the **Color** of the light.

- Under **Intensity**, change the light's brightness.

- Under **Rotation**, adjust the angled direction of the light.

- Under **Height**, adjust the distance of the light from the object, which affects the length of the shadows.

- Under **Softness**, adjust the spread of the light.

- Select or deselect **Ambient Light** to determine whether there is visible global lighting and to modify its **Intensity**.

- Under **Shadows**, determine whether shadows are applied to the effect.

- Under **Position**, determine whether the shadow is behind or below the object.

- Under **Distance from Object**, adjust the space between the object and the shadow.

- Under **Shadow Bounds**, adjust the size of the shadow's boundary.

Manually adjust an object's rotation using the 3D widget

The 3D widget appears for selected objects that have an effect applied using the **3D and Materials** panel (**Figure 15.5**).

With the 3D object selected, do any of the following:

- Click+drag the widget's horizontal bar up or down to rotate the object around its x axis (**Figure 15.6**).

- Click+drag the widget's vertical bar right or left to rotate the object around its y axis (**Figure 15.7**).

- Click+drag the widget's outer circle to rotate the object around its z axis (**Figure 15.8**).

- Click+drag the widget's center circle to rotate the object freeform (**Figure 15.9**).

FIGURE 15.5
The 3D widget visible with a selected 3D object

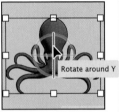 

FIGURE 15.6 Rotating a 3D object around its x axis

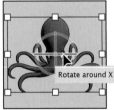 

FIGURE 15.7 Rotating a 3D object around its y axis

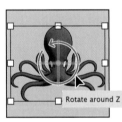 

FIGURE 15.8 Rotating a 3D object around its z axis

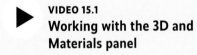

VIDEO 15.1
Working with the 3D and Materials panel

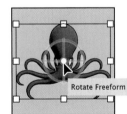 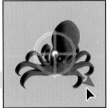

FIGURE 15.9 Rotating a 3D object freeform around its center

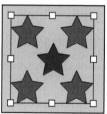

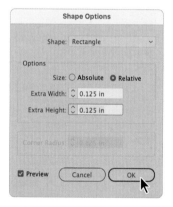

FIGURE 15.10 Applying an ellipse shape effect using absolute dimensions

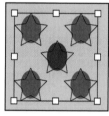

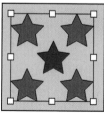

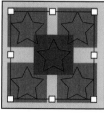

FIGURE 15.11 Applying a rectangle shape effect using relative dimensions

Convert one shape into another

The **Convert to Shape** effects let you experiment with changing objects into basic shapes or swapping one shape for another. This can be particularly helpful when creating patterns.

With the objects selected, do the following:

1. Choose **Effect** > **Convert to Shape** > *[shape effect name]*.

2. In the **Shape Options** dialog box, choose either **Absolute** and size the objects using their own dimensions (**Figure 15.10**) or choose **Relative** and size the objects using the original object's boundaries (**Figure 15.11**).

3. Click OK.

Apply crop marks to objects

To apply crop marks as an effect, do the following (**Figure 15.12**):

1. Select the object or group of objects.

2. Choose **Effect** > **Crop Marks**.

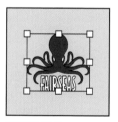

FIGURE 15.12 Applying crop marks to a group of objects

> **TIP** Unless you want to apply individual crop marks to each selected object, be sure to group them before applying the effect.

Apply Distort & Transform effects to objects

Distort & Transform effects (**Figure 15.13**) let you easily reshape object appearances to create visual interest.

With the objects selected, do the following:

1. From the **Effect** menu or **fx** button, choose **Distort & Transform** > *[effect name]*.

2. In the corresponding dialog box, customize the effect settings as needed, and then click **OK**.

Original

Free Distort

Pucker

Bloat

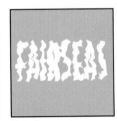
Roughen

Transform

Tweak

Twist

Zig Zag

FIGURE 15.13 Distort & Transform effect examples

Apply the Offset Path effect to objects

With the objects selected, do the following (**Figure 15.14**):

1. From the **Effect** menu or **fx** button, choose **Path > Offset Path**.

2. In the **Offset Path** dialog box, customize the settings as needed, and then click **OK**.

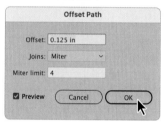

FIGURE 15.14 Applying the Offset Path effect to selected objects

Apply the Outline Object effect to text

Applying the **Outline Object** effect to text objects lets you treat them as path objects while still retaining their text properties.

With the text objects selected, do the following:

- From the **Effect** menu or **fx** button, choose **Path > Outline Object**.

Apply the Outline Stroke effect to an object

Applying the **Outline Stroke** effect to objects lets you treat their strokes as shapes while still retaining their stroke properties.

With the stroked objects selected, do the following:

- From the **Effect** menu or **fx** button, choose **Path > Outline Stroke**.

Apply Pathfinder effects to objects

Applying Pathfinder effects to objects lets you change the appearance of objects by combining them into new shapes or dividing them into separate ones, while keeping the original objects intact.

With the objects selected, do the following:

- From the **Effect** menu or **fx** button, choose **Distort & Transform** > *[effect name]*.

TIP Pathfinder effects are an advanced feature. In most cases, using the Pathfinder panel is more appropriate. For descriptions of the individual Pathfinder effects, see the "Pathfinder panel actions defined" sidebar of the "Modifying Objects Using Panels" section in Chapter 13.

Apply the Rasterize effect to objects

The **Rasterize** effect lets you preview how your vector artwork will appear when converted to bitmap images, while retaining their vector properties.

With the objects selected, do the following (**Figure 15.15**):

1. From the **Effect** menu or **fx** button, choose **Rasterize**.

2. In the **Rasterize** dialog box, customize the settings as needed, and then click **OK**.

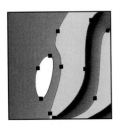 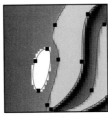

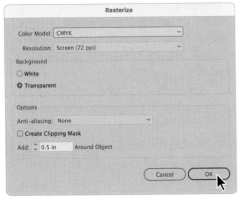

FIGURE 15.15 Applying the Rasterize effect to selected objects

TIP When rasterizing artwork with multiple overlapping paths, be sure to group them before applying the effect.

Rasterize setting options

Change the **Color Model** if you will be using the rasterized artwork for a different purpose than the original (color to grayscale or bitmap).

Adjust the **Resolution** to determine the number of pixels per inch (ppi).

Choose whether the **Background** of the rasterized area will be **White** or **Transparent**.

Choose an **Anti-aliasing** option depending on how you want the pixels to appear:

- **None** produces visible pixel steps with no smoothing.

- **Art Optimized** applies the most appropriate anti-aliasing for artwork with no text elements.

- **Type Optimized** applies the most appropriate anti-aliasing for artwork that contains text elements.

Choose **Create Clipping Mask** if you want to apply a smooth vector edge to the outside parts of the original vector paths as a clipping mask to the pixel element.

Adjust the **Add Around Object** value to include padding for the rasterized image.

Original

Drop Shadow

Feather

Inner Glow

Outer Glow

Round Corners

Scribble

FIGURE 15.16 Vector Stylize effect examples

Apply Illustrator Stylize effects to objects

Stylize effects (**Figure 15.16**) let you easily add depth, lighting, and art appearances to your objects.

With the objects selected, do the following:

1. From the **Effect** menu or **fx** button, choose **Stylize** > *[effect name]*.

2. In the corresponding dialog box, customize the effect settings as needed, and then click **OK**.

Illustrator Stylize setting options

Choose a **Mode** to set the blending mode for drop shadows and glow effects.

Adjust the **Opacity** to determine the transparency of the shadow or glow effect.

Adjust the **X Offset and Y Offset** to determine the horizontal and vertical distance of the shadow effect from the object.

Adjust the **Blur** to determine the distance from the edge of the shadow or glow where the blurring should occur.

Adjust the **Color** of the shadow or glow.

Adjust the **Darkness** to set the level of black applied to the shadow.

Adjust the size of the **Radius** for the **Feather** or **Round Corners** effects.

For the **Scribble** effect, choose a preset **Setting** and/or customize the **Angle**, **Path Overlap**, **Variation**, and **Line Options** for the effect display behavior.

Apply SVG Filter effects to objects

SVG Filter (Scalable Vector Graphic) effects (**Figure 15.17**) are resolution-independent, XML-based properties that can be used when you export SVG files for web-based purposes. The effects will be exported as code for the browser to render.

With the objects selected, do the following:

- From the **Effect** menu or **fx** button, choose **SVG Filter** > *[effect name]*.

Original

Alpha

BevelShadow

CoolBreeze

Dilate

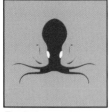
Erode

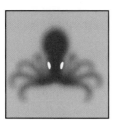
GaussianBlur

PixelPlay

Shadow

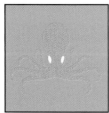
Static

Turbulence

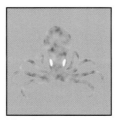
Woodgrain

FIGURE 15.17 SVG Filters effect examples

TIP You can also import SVG Filter effects by choosing Import SVG Filter. Additionally, you can edit or create your own, by choosing Apply SVG filter and applying XML code.

Original

Apply Warp effects to objects

Warp effects (**Figure 15.18**) create visual interest by distorting and deforming the appearance of artwork using customizable preset warp shapes.

With the objects selected, do the following:

1. From the **Effect** menu or **fx** button, choose **Warp** > *[effect name]*.

2. In the **Warp** dialog box, customize the settings as needed, and then click **OK**.

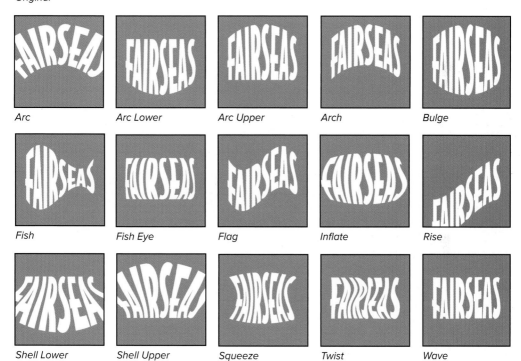

Arc   Arc Lower   Arc Upper   Arch   Bulge

Fish   Fish Eye   Flag   Inflate   Rise

Shell Lower   Shell Upper   Squeeze   Twist   Wave

FIGURE 15.18 Warp effect examples

TIP In addition to objects and text, Warp effects can be applied to raster images, blends, and meshes.

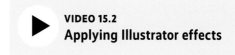
VIDEO 15.2
Applying Illustrator effects

Applying Photoshop Effects

Photoshop effects let you add traditional art appearances or lighting and distortion effects.

Apply a raster effect using the Effect Gallery dialog box

In the **Effect Gallery** dialog box, do the following:

1. Either click an effect thumbnail (**C** in Figure 15.19) in one of the categories or choose an option from the **Effect** menu (**D** in Figure 15.19).

2. (Optional) Adjust the settings for the selected effect options (**E** in Figure 15.19).

Access the Effect Gallery dialog box

Most Photoshop effects (with the exceptions of Blur, Pixelate, and Video) are included in the **Effect Gallery** dialog box (**Figure 15.19**), which provides an efficient visual workspace for experimenting with various raster effects.

Do the following:

1. Select the objects for applying the effect, and group them (**Object** > **Group**).

2. From the **Effect** menu or **fx** button, choose **Effect Gallery**.

> **TIP** Selecting an effect included in the Effect Gallery using the Effect menu will automatically open the Effect Gallery dialog box.

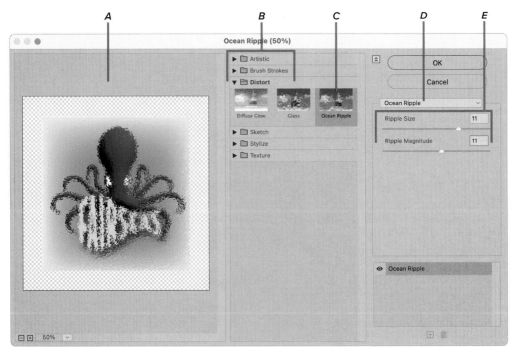

FIGURE 15.19
A. Preview area **B.** Effect categories **C.** Selected effect thumbnail **D.** Effect selection menu
E. Selected effect options

Artistic Effects

Artistic effects (**Figure 15.20**) replicate the appearance of traditional art media.

- **Colored Pencil** applies a cross-hatch appearance over a solid background, retaining important edges.
- **Cutout** creates the appearance of roughly cut colored paper for the image elements.
- **Dry Brush** replicates the traditional painting technique, reducing the number of colors, which simplifies the image.
- **Film Grain** unifies the element appearances and eliminates banding by applying even patterns to the image's mid and shadow tones, and applying more saturation patterns to the lighter areas.
- **Fresco** replicates the traditional coarse painting technique using hastily applied short, rounded strokes.
- **Neon Glow** softens the look of the image and colorizes it by adding different glow types to the various elements.
- **Paint Daubs** replicates the traditional painting technique, letting you customize the brush type and size.
- **Pallet Knife** replicates the traditional painting technique, which reduces the image's detail.

FIGURE 15.20 Artistic effects in the Effect Gallery dialog box

- **Plastic Wrap** creates the appearance of applying a transparent shiny covering over the image, which emphasizes the surface details.
- **Poster Edges** applies black lines to the element edges and reduces the number of colors in the image, dependent on the values you choose.
- **Rough Pastels** replicates the traditional colored chalk technique, using strokes on a textured background.
- **Smudge Stick** replicates the traditional art technique by softening the image by smearing the darker areas using short diagonal strokes, and brightening and diffusing the lighter areas.
- **Sponge** replicates the traditional painting technique by using high amounts of texture in areas where the image has contrasting colors.
- **Underpainting** replicates the traditional painting technique of painting the image both on the textured background, and then again on top of it.
- **Watercolor** replicates the traditional painting technique, using liquid brush appearances that simplify the image details and saturating the color along edges of tonal contrast.

Brush Stroke Effects

Brush Stroke effects (**Figure 15.21**) replicate traditional ink, brush, and drawing material visuals.

- **Accented Edges** emphasizes the edges of elements in the image. High values applied to the Edge Brightness replicate white chalk, while lower values replicate black ink.

- **Angled Strokes** applies diagonal strokes to repaint the image, using a single direction for lighter areas, and adding strokes in the opposite direction for darker areas.

- **Crosshatch** leaves the original image intact, simulating pencil hatching by adding roughness to the edges and texture to the colored areas.

- **Dark Strokes** applies shorter strokes to the dark areas and longer strokes to the lighter areas.

- **Ink Outlines** applies narrow pen-and-ink fine lines over the original image.

FIGURE 15.21 Brush Stroke effects in the Effect Gallery dialog box

- **Spatter** replicates an airbrush's spatters. The image detail decreases when higher values are applied.

- **Sprayed Strokes** replicates an airbrush's angled strokes using the dominant colors.

- **Sumi-e** replicates the traditional Japanese ink-on-rice paper technique using a wet brush, resulting in rich blacks with softened edges.

Distort Effects

Distort effects (**Figure 15.22**) use geometric configurations to reshape and distort the image.

- **Diffuse Glow** replicates using a soft diffusion filter to view the image, with a glow that fades from the center of a selected area.

- **Glass** replicates using different glass types to view the image, by using provided presets or creating your own.

FIGURE 15.22 Distort effects in the Effect Gallery dialog box

- **Ocean Ripple** replicates looking at the image as if it were under water, by applying random ripples to the artwork.

Sketch Effects

Sketch effects (**Figure 15.23**) replicate traditional hand-drawn techniques that also add texture to the appearance.

- **Bas Relief** replicates low-relief carving, which accents variations in the image surfaces.
- **Chalk & Charcoal** replicates these drawing materials using the appearance of coarse chalk strokes for the light areas, diagonal charcoal strokes for the dark areas, and solid gray for the midtones.
- **Charcoal** replicates this drawing material to apply a smudged, posterized effect to the image.
- **Chrome** alters the image to appear as though it is a reflection on a highly polished metallic surface.
- **Conté Crayon** replicates this drawing material using black for the dark areas and white for the light areas.
- **Graphic Pen** replicates this drawing material using narrow pen-and-ink fine strokes to replace the image details, using black ink for the dark areas and the white paper for light areas.
- **Halftone Pattern** maintains the image's continuous tones while applying a halftone screen appearance.
- **Note Paper** replicates using layers of cut textured paper for the image elements, creating an embossed effect. Dark areas on the top layer appear as holes in the paper.
- **Photocopy** creates the appearance of a black-and-white photocopied image. Large areas of darkness tend to copy only around their edges.
- **Plaster** creates the appearance of the image as black-and-white and molded in plaster.
- **Reticulation** replicates stippling and creates the appearance of film emulsion's controlled shrinkage and distortion. This results in the shadow areas appearing clumped and the highlight areas appearing lightly grained.
- **Stamp** replicates a rubber or wood stamp impression of the image.
- **Torn Edges** creates the appearance of raggedly torn black-and-white pieces of paper for the image elements.
- **Water Paper** creates the appearance of damp textured paper with blotchy paint daubs of color that blend and flow.

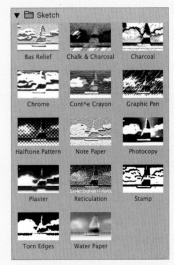

FIGURE 15.23 Sketch effects in the Effect Gallery dialog box

Photoshop Stylize Effect

The Photoshop **Stylize** effect category (**Figure 15.24**) contains the **Glowing Edges** effect, which displaces pixels and increases the image contrast, resulting in an impressionistic or glowing appearance.

FIGURE 15.24 The Glowing Edges effect in the Stylize category of the Effect Gallery dialog box

Texture Effects

Texture effects (**Figure 15.25**) add the appearance of depth, organic qualities, or substance to the artwork.

- **Craquelure** creates the appearance of the image painted onto a high-relief plaster surface, resulting in a delicate network of cracks relative to the image's contours.

- **Grain** simulates various types of granulation (regular, soft, sprinkles, clumped, contrasty, enlarged, stippled, horizontal, vertical, or speckle).

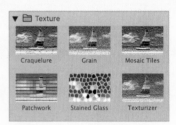

FIGURE 15.25 Texture effects in the Effect Gallery dialog box

- **Mosaic Tiles** creates the appearance of the image being composed of small tiles or chips with grout between them.

- **Patchwork** divides the image into single-color adjacent squares using the image's predominant color within that area.

- **Stained Glass** divides the image into single-colored adjacent cells using the image's predominant color within that area, and outlines them using the foreground color.

- **Texturizer** adds a textural appearance to the image using provided presets or by creating your own in Photoshop.

Blur settings

Gaussian Blur adds the simulation of haziness.

Radial Blur simulates camera lens effects:

- **Spin** blurs using concentric circular paths and specified rotation.

- **Zoom** blurs radially as when zooming in and out.

- **Blur Center** box lets you set the blur origin by dragging the pattern.

Smart Blur lets you precisely determine the blurring appearance:

- **Radius** sets what size of area is searched for dissimilar pixels.

- **Threshold** sets how different the pixels need to be in order to be included.

- **Normal** blurs the entire selection.

- **Edge Only** and **Overlay** blur only the color transition edges.

TIP When applying blur effects to multiple objects, be sure to group them first.

Apply a Blur effect

Blur effects (**Figure 15.26**) let you add the appearance of haziness or motion to artwork.

With the objects selected, do the following:

1. From the **Effect** menu or **fx** button, choose **Blur** > *[effect name]*.

2. In the corresponding dialog box, customize the settings as needed, and then click **OK**.

Original

Gaussian Blur

Radial Blur

Smart Blur

FIGURE 15.26 Raster Blur effect examples

VIDEO 15.3
Applying Photoshop effects

Apply a Pixelate effect

Pixelate effects (**Figure 15.27**) clump similar adjacent pixels with similar color values to add sharp definitions to the artwork.

With the objects selected, do the following:

1. From the **Effect** menu or **fx** button, choose **Pixelate**> *[effect name]*.

2. In the corresponding dialog box, customize the settings as needed, and then click **OK**.

Apply a Video effect

Video effects let you optimize the appearance of images captured from digital video sources.

With the objects selected, do the following:

1. From the **Effect** menu or **fx** button, choose **Video** > *[effect name]*.

2. In the corresponding dialog box, customize the settings as needed, and then click **OK**.

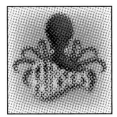
Color Halftone

Crystallize

Mezzotint

Pointillize

FIGURE 15.27 Pixelate effect examples

Video effects

De-Interlace removes interlaced lines from the image to smooth the appearance of the video. The discarded lines can be replaced by duplication or interpolation.

NTSC Colors prevents oversaturated colors from bleeding across TV scan lines by restricting the color gamuts to those acceptable (NTSC) for TV production.

Pixelate effects

Color Halftone divides each color channel into rectangles and then replaces those rectangles with circles that are sized proportionally to their brightness.

- *Grayscale* images use **Channel 1**.

- *RGB* images use **Channels 1–3**.

- *CMYK* images use **Channels 1–4**.

Crystallize uses polygon shapes to clump the colors.

Mezzotint creates the appearance of randomly patterned areas of black and white or fully saturated colors. (For **Type**, a **Dot** pattern must be the selected to apply to saturated colors.)

Pointillize replicates the appearance of pointillist painting by dividing the image colors into randomly placed dots and using the background color as the canvas area between them.

Managing Appearance Attributes

The Properties and Appearance panels help you modify artwork appearance attributes easily. Saving applied attributes as graphic styles lets you apply them to other elements efficiently.

In This Chapter

Modifying Applied Effects Using the Properties Panel

In addition to fill and stroke, the **Properties** panel lets you access singularly applied effects. However, accessing multiple applied effects must be done using the **Appearance** panel.

Use the Properties panel to modify or delete an effect

With the objects selected and the **Properties** panel active, do any of the following:

- Click the effect name to open the corresponding dialog box or panel to edit the effect (**Figure 16.1**).

- Click the **Delete Effect** icon next to the effect name to remove the effect (**Figure 16.2**).

- If a circle with the letter "i" icon is visible (**Figure 16.3**), this means that there are multiple effects applied to the selection. Open the **Appearance** panel to modify or delete them.

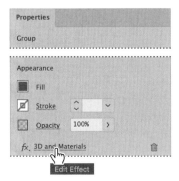

FIGURE 16.1 Clicking an applied effect name to modify its settings

> **TIP** Once applied, effects cannot be edited by choosing them from the Effect menu or fx button. Using the Effect menu or fx button will apply a duplicate of the effect to the selection.

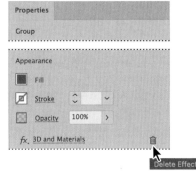

FIGURE 16.2 Removing an applied effect

FIGURE 16.3 A selected group with multiple effects applied

Using the Appearance Panel

The **Appearance** panel (**Figure 16.4**) uses a hierarchical structure to help you view and modify a selection's applied attributes and effects.

Access the Appearance panel

Do either of the following:

- Click the **More** (...) button in the **Appearance** section of the **Properties** panel (**Figure 16.5**).

- Choose **Window** > **Appearance**.

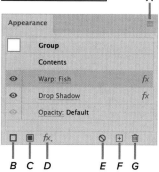

FIGURE 16.4
A. Panel menu **B.** Add New Stroke
C. Add New Fill **D.** Add New Effect
E. Clear Appearance **F.** Duplicate
Selected Item **G.** Delete Selected Item

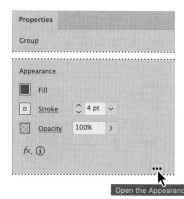

FIGURE 16.5 Clicking More button to open the Appearance panelfrom the Properties panel

View additional applied attributes

If the selection contains groups or multiple layers, the **Appearance** panel displays a **Contents** row, rather than all of the selection's attributes.

In the **Appearance** panel, do the following (**Figure 16.6**):

- Double-click the **Contents** row to view the additional attributes.

TIP To view the applied effects again, click the top-level row.

FIGURE 16.6 Viewing additional attributes in the Appearance panel

Use the Appearance panel to modify an applied attribute or effect

With the objects selected and the **Appearance** panel active, do any of the following:

- Click the underlined effect or attribute name to open the corresponding dialog box or panel.

- Double-click the effect or attribute row to open the corresponding dialog box or panel.

Show or hide an attribute or effect

In the **Appearance** panel, do the following (**Figure 16.7**):

- Click the **eye** icon for the item.

FIGURE 16.7 Hiding an applied effect

Use the Appearance panel to delete an applied attribute or effect

With the objects selected and the **Appearance** panel active, do any of the following:

1. Click to select the effect or attribute.

2. Click the **Delete Selected Item** icon (**G** in Figure 16.4) or choose **Remove Item** from the panel menu.

Remove all appearance attributes or effects

With the objects selected and the **Appearance** panel active, do either of the following:

- Click the **Clear Appearance** button (**Figure 16.8**).

- Choose **Clear Appearance** from the panel menu.

FIGURE 16.8 Removing applied effects from a group

TIP If you want to also remove all the fill and stroke attributes from the selection, access them by double-clicking the Contents row.

Add an attribute or effect

With the effect or attribute selected that you want to place the new effect or attribute above, in the **Appearance** panel, do any of the following:

- Apply a new stroke by clicking the **Add New Stroke** button (**B** in Figure 16.4) or choosing Add New Stroke from the panel menu (**Figure 16.9**).

- Apply a new fill by clicking the **Add New Fill** button (**C** in Figure 16.4) or choosing **Add New Fill** from the panel menu.

- Apply a new effect by clicking the **Add New Effect** button (**D** in Figure 16.4) or choosing **Add New Effect** from the panel menu.

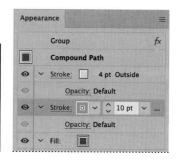

FIGURE 16.9 Adding and customizing a new stroke attribute between the existing stroke and fill

Reorganize attributes or effects

In the **Appearance** panel, with the effect or attribute you want to move selected, do the following (**Figure 16.10**):

- Click+drag the item to the desired location.

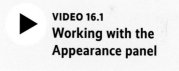

VIDEO 16.1
Working with the Appearance panel

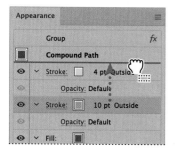

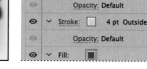

FIGURE 16.10 Moving a new stroke above the original one

Working with Graphic Styles

The **Graphic Styles** panel (**Figure 16.11**) lets you quickly apply attributes to objects or save them for use with other elements.

Access the Graphic Styles panel

Do either of the following:

- Choose **Window** > **Graphic Styles**.
- In the **Essentials Classic** workspace, click the **Graphic Styles** thumbnail (**Figure 16.12**).

Apply a graphic style

With the objects you want to apply the graphic style to selected, do the following:

- In the **Graphic Styles** panel, click the style thumbnail.

Remove an applied graphic style

With the objects you want to remove the graphic style from selected, do either of the following:

- In the **Graphic Styles** panel, click the **Default Graphic Style** thumbnail (**A** in Figure 16.11) to apply a black stroke and white fill.
- In the **Appearance** panel, click the **Clear Appearance** button (**Figure 16.13**) to delete all attributes and effects, making it invisible.

Delete a graphic style from the panel

With the style thumbnail selected in the **Graphic Styles** panel, do either of the following:

- Click the **Delete Graphic Style** button (**G** in Figure 16.11).
- Choose **Delete Graphic Style** from the panel menu.

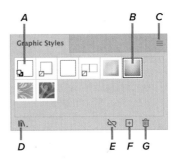

FIGURE 16.11
A. Default Graphic Style **B.** Applied Graphic Style
C. Panel menu **D.** Graphic Styles Libraries menu
E. Break Link to Graphic Style **F.** New Graphic Style
G. Delete Graphic Style

FIGURE 16.12 Opening the Graphic Styles panel from its thumbnail

FIGURE 16.13 Using the Appearance panel to remove an applied graphic style from an object

Create a graphic style

With the object you want to create the graphic style from selected, do the following:

1. In the **Appearance** panel, choose whether you want to create the style from the applied effects or attributes (under **Contents**).

2. In the **Graphic Styles** panel, do either of the following:

   Click the **New Graphic Style** button (**F** in Figure 16.11) to add the attributes as an unnamed style.

   Choose **New Graphic Style** from the panel menu, and then use the **Graphic Style Options** dialog box to enter a **Style Name** (Figure 16.14).

TIP When you break a selection's link to a graphic style, it retains all the applied style attributes except those you modify.

Rename a graphic style

In the **Graphic Styles** panel, do either of the following to open the **Graphic Style Options** dialog box and rename the style:

- Double-click the style thumbnail.

- Select the style thumbnail, and then choose **Graphic Style Options** from the panel menu.

TIP When organizing graphic styles, make sure no objects are selected so you do not inadvertently apply a style to them.

Break an applied link to a graphic style

With the objects with the applied style selected, do either of the following:

- In the **Graphic Styles** panel, click the **Break Link to Graphic Style** button (**E** in Figure 16.11) or choose **Break Link to Graphic Style** from the panel menu.

- Change any of the selection's appearance attributes (fill, stroke, effect, etc.).

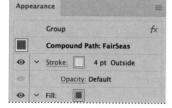

TIP When you create a graphic style from a selection's attributes, it is automatically applied in the Appearance panel as well.

FIGURE 16.14 Creating a new named graphic style

Using Graphic Style Libraries

Graphic Style Libraries contain collections of preset graphic styles that open in their own panels.

Open a graphic style library

Do either of the following:

- In the **Graphic Styles** panel, choose a library by clicking the **Graphic Styles Libraries** button (**Figure 16.15**) or by choosing **Open Graphic Style Library** from the panel menu.

- Choose **Window** > **Graphic Style Libraries** > *[library name]*.

TIP Graphic style library panel items can be applied and arranged, but cannot be modified or deleted.

Create a graphic style library

In the **Graphic Styles** panel, do the following:

1. Organize the graphic styles as needed.

2. Choose **Save Graphic Styles** from the **Graphic Styles Libraries** button menu, or choose **Save Graphic Style Library** from the panel menu.

TIP Saving the library file to the default location allows it to appear under the User Defined submenu of the Graphic Style Libraries menu.

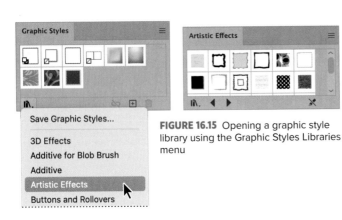

FIGURE 16.15 Opening a graphic style library using the Graphic Styles Libraries menu

Add a library's graphic style to the Graphic Styles panel

Do either of the following:

- With the style thumbnail selected in its library panel, drag it into the **Graphic Styles** panel, or choose **Add to Graphic Styles** from the panel menu.

- Apply the style to an object to automatically add it to the **Graphic Styles** panel.

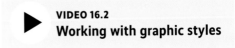

VIDEO 16.2
Working with graphic styles

Importing Assets

Illustrator recognizes all common vector and raster file formats, allowing you to easily incorporate other artwork and images into your document's design.

Placing Files

Using the **Place** commands to import artwork provides the most functionality and support when importing documents.

Place a linked file

Do the following (**Figure 17.1**):

1. Choose **File** > **Place** to open the dialog box and select the file.

2. Select the **Link** option.

3. Click **Place**.

4. In the document window, position the cursor where you want to place the upper-left corner of the file.

5. Click to place the file.

> **TIP** Linked placed images display a cross-hatch when selected.

Linking vs. embedding files

When you place artwork or image files, you can choose whether to link or embed them.

Linked files are connected to the document, but also independent from it. This preserves the original and reduces file size. Linked files can be edited only by modifying the source document.

Embedded files are complete editable copies of the source document. Embedded files will increase your document's size.

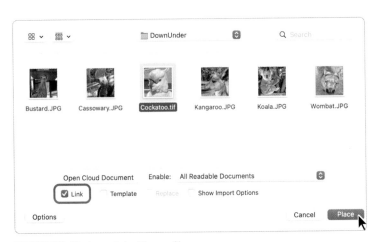

FIGURE 17.1 Placing a linked image file

> **TIP** Illustrator also lets you choose and place multiple files one at a time using Shift or Command/Control to select them in the dialog box. The cursor will display how many files are loaded and will place them sequentially as you click in the document window.

> **TIP** You can also place a linked file by dragging it from your computer's desktop window onto the Illustrator document window.

Place an embedded file

Do the following (**Figure 17.2**):

1. Choose **File** > **Place** to open the dialog box and select the file.

2. Deselect the **Link** option.

3. Click **Place**.

4. In the document window, position the cursor where you want to place the upper-left corner of the file.

5. Click to place the file.

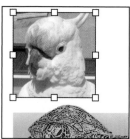

FIGURE 17.2 Placing an embedded image file

> **TIP** You can also add embedded elements by using the Copy/Cut and Paste commands or by pressing Shift as you drag it from your computer's desktop window onto the Illustrator document window.

Open an image file as an Illustrator document

When you open an image file directly in Illustrator, the image is placed as an *embedded* file using the most recently chosen Illustrator artboard dimensions (**Figure 17.3**).

To open an image file from the **Home** screen, do any of the following:

- Click the **Open** button and then navigate to select the image file.

- Click one of the options under **Your Work** and select the image file.

- Click an image file listed under the **Recent** section.

To open an image file from the **Application** frame, do any of the following:

- Choose **File** > **Open** and navigate to the image file.

- Choose **File** > **Open Recent Files** and select a recently opened image file from the context menu.

FIGURE 17.3 Opening an image file in Illustrator

Managing Placed Files

The **Links** panel (**Figure 17.4**) provides access to, and information about, all the placed artwork files in your document.

The **Object Type** and **Links** sections of the **Control** panel (**Figure 17.5**) let you manage the settings for a selected placed artwork file.

Open the Links panel

Do either of the following:

- Choose **Window** > **Links**.
- Click the underlined **Object Type** name in the **Control** panel (Figure 17.5).

View a placed file's information

Do the following (**Figure 17.6**):

- Click the **Show Link Info** button at the bottom of the **Links** panel.

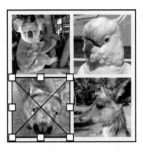

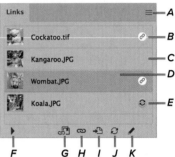

FIGURE 17.4
A. Panel menu **B.** Linked file
C. Embedded file **D.** Selected file
E. Modified linked file **F.** Show Link Info
G. Relink from CC Libraries **H.** Relink
I. Go to Link **J.** Update Link
K. Edit Original or Edit in Photoshop

FIGURE 17.5 The Object Type (Linked File) and Link section of the Control panel for the selected file

TIP Basic link commands are also available in the Quick Actions section of the Properties panel.

View a placed file's metadata

Metadata is additional information (copyright information, descriptions, etc.) that can be added to the source file.

Do the following:

- Choose **Link File Info** from the **Links** panel menu.

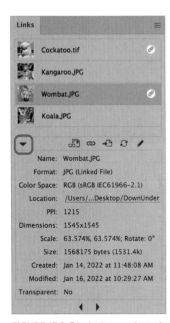

FIGURE 17.6 Displaying a selected file's link information

Embed a linked file

With the linked file selected, do any of the following:

- Click **Embed** in the **Links** section of the **Control** panel.

- Click **Embed** in the **Quick Actions** section of the **Properties** panel.

- Choose **Embed Image(s)** from the **Links** panel menu.

Link an embedded file

With the embedded file selected, do the following:

1. Click **Unembed** in the **Control** or **Properties** panel, or choose **Unembed** from the **Links** panel menu.

2. In the **Unembed** dialog box, choose a file format and then save the file to the desired location.

Update a modified linked file

If the source file has been modified after placing it in your document (**E** in Figure 17.4), do either of the following in the **Links** panel:

- Click the **Update Link** button (**I** in Figure 17.4).

- Choose **Update Link** from the panel menu.

TIP Illustrator will also alert you with a warning dialog box when a file has been modified outside of Illustrator, allowing you to update it by clicking Yes.

Replace a placed file

With the file to be replaced selected, do any of the following (**Figure 17.7**):

- Choose **File** > **Place** and with the replacement file chosen, select the **Replace** option.

- In the **Links** panel, click the **Relink** button (**H** in Figure 17.4) or choose **Relink** from the panel menu to select a replacement file from the dialog box.

FIGURE 17.7 Replacing an embedded image file with a different linked file

Modify a linked source file

Do any of the following

- In the **Links** panel, click the **Edit Original** button (**K** in Figure 17.4) or choose **Edit Original** from the panel menu.

- In the **Control** or **Properties** panel, click **Edit Original**.

TIP If you have Photoshop installed, linked image files will display Edit in Photoshop instead of Edit Original. More information is available in the "Importing Photoshop (.psd) Files" section of this chapter.

Using Image Trace to Convert Raster Images to Vector Artwork

The **Image Trace** panel (**Figure 17.8**) lets you easily convert digital or scanned images and drawings into vector artwork, using either presets or a robust collection of customizable settings.

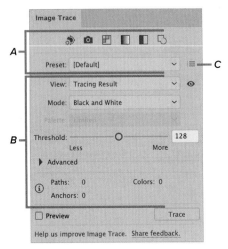

FIGURE 17.8
A. Preset buttons and menu
B. Customizable settings
C. Manage Presets button

Apply Image Trace using a preset

With the image selected, do either of the following (**Figure 17.9**):

- In the **Image Trace** panel, click a preset button or choose a setting from the **Preset** menu.
- In the **Control** panel, choose a preset from the **Image Trace** menu button (**Figure 17.10**).

Access the Image Trace panel

Do either of the following:

- Choose **Window** > **Image Trace**.
- Switch to the **Tracing** workspace and then select the **Image Trace** panel tab.

Original

Low Color   *Grayscale*   *Black and White*

6 Colors   *3 Colors*   *Silhouettes*

FIGURE 17.9 Examples of Image Trace presets

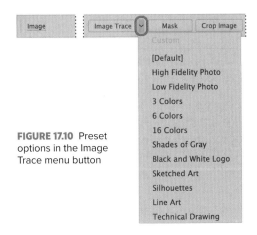

FIGURE 17.10 Preset options in the Image Trace menu button

Apply Image Trace using customized options

With the image selected and the **Image Trace** panel active, do any of the following:

- Choose a **View** to determine how the traced object is displayed.

- Choose a **Mode** to determine the color mode for the tracing result.

- If a color mode is selected, choose a **Palette** option for generating the tracing.

- Dive deeper by investigating additional options under the **Advanced** button.

Apply Image Trace using the default (black and white) setting

With the image selected, do either of the following:

- Choose **Object** > **Image Trace** > **Make**.

- Click the **Image Trace** button in either the **Control** or **Properties** panel.

Save customized tracing settings as a preset

With the customized settings active in the **Image Trace** panel, do the following:

1. In the **Preset** section, click the **Manage Presets** button (**C** in Figure 17.8) and choose **Save as New Preset**.

2. In the **Save Image Trace Preset** dialog box, enter a name and click **OK**.

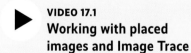

VIDEO 17.1
Working with placed images and Image Trace

Display the traced outlines

Do the following:

- Under **View** in the **Image Trace** panel, choose an outline display option (**Figure 17.11**).

Tracing Result (default) *Tracing Result with Outlines*

Outlines *Outlines with Source Image*

FIGURE 17.11 Traced outline views

Convert the tracing results to paths

Converting the tracing results is the important final step in letting you work with them as vector objects.

With the result selected, do the following:

1. Click **Expand** in the **Control** or **Properties** panel, or choose **Object** > **Image Trace** > **Expand**.

2. Ungroup the paths by clicking **Ungroup** in the **Properties** panel or choosing **Object** > **Ungroup**.

Importing Photoshop (.psd) Files

When you open or place an embedded Photoshop file into Illustrator, you have the option of preserving many of Photoshop's features, including layers, text, and paths (**Figure 17.12**).

Access the Photoshop Import Options dialog box

Do either of the following:

- Place the .psd file as embedded (**Link** deselected) with **Show Options** selected (**Figure 17.13**).

- Open the .psd file in Illustrator (**Figure 17.14**).

FIGURE 17.12 Preserving the Photoshop layer structure in Illustrator

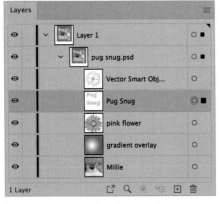

FIGURE 17.13 Photoshop file placed in Illustrator with text selected

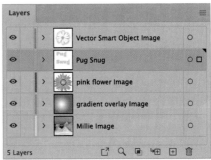

FIGURE 17.14 Photoshop file opened in Illustrator with text selected

Importing Acrobat (.pdf) Files

Illustrator lets you determine which pages of a PDF document are imported when you open or place them. Placing a file lets you choose a single page, while opening a PDF file provides the option to choose multiple pages that open on their own artboards.

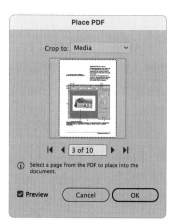

FIGURE 17.15 Selecting a PDF page to place

Place a PDF file using page options

Do the following:

1. When placing the PDF file, select **Show Options** in the dialog box and then click **Place.**

2. In the **Place PDF** dialog box, choose the source you want to **Crop To** and which PDF page to place (**Figure 17.15**).

3. Click **OK**.

Open a PDF file using page options

Select the file to open in Illustrator, and then in the **PDF Import Options** dialog box, do any of the following (**Figure 17.16**):

-   View the page thumbnails using the navigational tools.

-   Choose which pages to open using the **Range** options.

-   Open every page by selecting **All**.

-   Open the pages as linked files by selecting **Import PDF pages as links for optimal performance**.

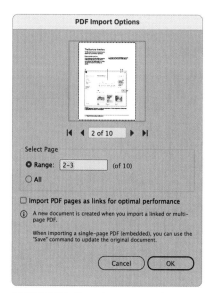

FIGURE 17.16 Choosing a range of pages to open in their own artboards

Importing Text Documents

Illustrator supports importing most Microsoft Word (.doc and .docx) formats, as well as Rich Text Format (.rtf) and plain text (.txt).

Open a text document in Illustrator

Do the following (**Figure 17.17**):

1. Choose **File** > **Open**.

2. In the **Open File** dialog box, select the text document and click **Open**.

3. (Optional) If you are opening a Word document, select the settings you want to customize, and then click **OK**.

FIGURE 17.17 Opening a Microsoft Word document in Illustrator

TIP For more information about work with text, see Chapter 11, "Adding and Customizing Text."

Place a text document in Illustrator

Do the following:

1. Choose **File** > **Place**.

2. In the **Place File** dialog box, select the text document and click **Open**.

3. (Optional) If you are opening a Word document, select the settings you want to customize, and then click **OK**.

TIP Placed text documents can only be embedded, not linked.

Saving and Exporting Files and Assets

Illustrator provides a number of tools to help ensure your saved and exported files are efficiently structured with the appropriate settings for the project type.

In This Chapter

Maximizing Document Efficiency

Prior to publishing or uploading a document, a number of tasks can be performed to maximize a document's efficiency.

Remove unneeded elements

Do any of the following:

- Delete unused layers, patterns, swatches, brushes, symbols, or styles.

- Delete hidden objects not needed for the document.

- Use **Object** > **Path** > **Cleanup** to delete stray points, unpainted objects, and empty text paths.

- Reduce file size by using Symbols for repeating elements. This allows Illustrator to use only one definition.

TIP Finalizing placed raster images in Photoshop prior to importing also helps with efficiency.

Rasterize complex vector artwork

Once you're sure your artwork is complete, rasterizing can help reduce the document size and avoid output display issues.

With the objects selected, choose **Object** > **Rasterize** and then do any of the following (**Figure 18.1**):

- Set the appropriate **Color Mode** and **Resolution**.

- Choose whether the **Background** should be **White** or **Transparent.**

- Choose the appropriate **Anti-aliasing** settings.

- Choose whether to preserve any spot colors.

TIP The Object > Rasterize command permanently converts the elements to pixels, removing all vector features. If you want to apply a raster appearance to vector objects without converting them, use Effect > Rasterize instead.

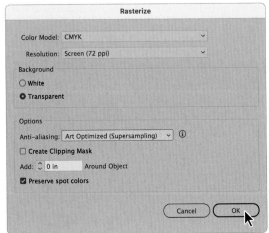

FIGURE 18.1 Rasterizing a complex group of vector objects

Managing Color Settings

While you determine your document's color mode when the file is created, you can modify the file settings for the document and also selected objects and embedded images.

Change the document's color mode

Do either of the following:

- Choose **File** > **Document Color Mode** > **RGB Color**.

- Choose **File** > **Document Color Mode** > **CMYK Color**.

Convert the color mode for selected elements

With the elements selected, do either of the following:

- Convert assigned spot colors to process colors by choosing **Edit** > **Edit Colors** > **Convert to CMYK** or **Convert to RGB**.

- Choose **Edit** > **Edit Colors** > **Convert to Grayscale** (**Figure 18.2**).

FIGURE 18.2 Converting selected color elements to grayscale

Modify Adobe color management settings

Adobe's color management system allows you to ensure color consistency by assigning profiles that decipher what an applied color was created by and which output intent the document has been designed for.

Do the following (**Figure 18.3**):

1. Choose **Edit** > **Color Settings**.

2. Select a preset from the **Settings** menu.

3. Click **Apply**.

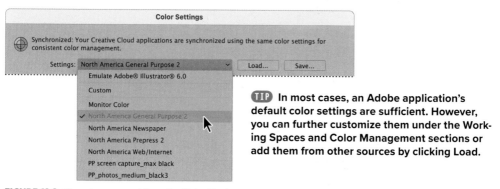

TIP In most cases, an Adobe application's default color settings are sufficient. However, you can further customize them under the Working Spaces and Color Management sections or add them from other sources by clicking Load.

FIGURE 18.3 Choosing a preset from the Color Settings menu

Display a document's proof settings

Do the following (**Figure 18.4**):

- Choose **View** > **Proof Colors**.

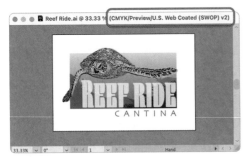

FIGURE 18.4 Displaying a document's proof color settings

Access soft color proofing options

Open the **Proof Setup** dialog box (**Figure 18.5**) by doing the following:

- Choose **View** > **Proof Setup** > **Customize**.

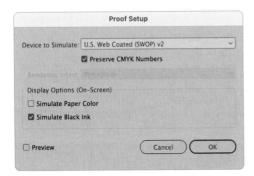

FIGURE 18.5 The Proof Setup dialog box

About color proofing

A **proof** is a simulation of the final output production process, such as a printing press. There are two proofing categories:

- **Hard proofs** are printed samples of the final product. These are usually produced on quality printers to provide an accurate representation.

- **Soft proofs** are onscreen simulations of the final product. These are often in PDF format. The quality of soft proofs is also dependent on the monitor it is viewed on.

Managing Metadata

Metadata is information you can add to your document such as authorship, description, and copyright information.

TIP Metadata fields can be modified in the Basic, Origin, IPTC, IPTC Extension, Audio Data, Video Data, and DICOM sections of the File Info dialog box.

Modify a document's metadata

Do the following (**Figure 18.6**):

1. Choose **File** > **File Info**.

2. Modify the fields as needed and then click **OK**.

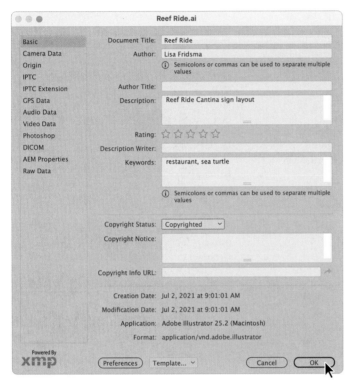

FIGURE 18.6 Customizing a document's metadata

Saving Illustrator Files

Saving your file in Illustrator (.ai) format helps ensure you retain all the document's features, such as layers and effects. However, Illustrator provides other format options as well.

Save an Illustrator file

Do any of the following:

- Choose **File** > **Save** to save a previously saved document using the default name and location.

- Choose **File** > **Save As** to open the **Save As** dialog box and assign a name and location for a new file (**Figure 18.7**).

- Choose **File** > **Save a Copy** to create a duplicate of the previously saved file. The original file remains the active open document.

Save an Illustrator file using an alternative format

Do the following:

- In the **Save As** or **Save a Copy** dialog box, choose a different option under the **Format** (macOS) or **Save as Type** (Windows) menu.

TIP Make sure you also save files using the most recent version of Illustrator to ensure you retain all their features.

Alternative formats for saving files

EPS (*Encapsulated PostScript*) preserves many of the elements created with Illustrator. EPS files can contain vector objects and raster images.

AIT (*Illustrator Template*) lets you create new documents with shared common settings and elements. When you create a new file based on a template, the original template file is preserved.

PDF (*Portable Document Format*) is an industry standard for document sharing. PDF files preserve the fonts, layout, images, and artwork for a wide array of source files.

SVG (*Scalable Vector Graphic*) is a scalable vector format often used for interactive and web applications.

SVGZ (*Compressed Scalable Vector Graphic*) reduces the file size of standard SVG format, but also removes the ability to modify the files using a text editor application.

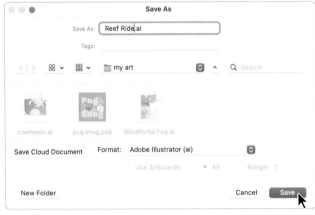

TIP Choosing File > Save for a previously unsaved document opens the Save As dialog box.

FIGURE 18.7 Saving an Illustrator file using the Save As dialog box

FIGURE 18.8 Creating a packaged folder and files

TIP If Collect Links in a Separate Folder is deselected, the asset copies are placed in the same folder with the Illustrator file.

TIP If Relink Linked files to the Document is deselected, the asset copies are still collected in the package folder, and the link path remains the same.

TIP Selecting Copy Fonts includes only the needed fonts, not the entire font family.

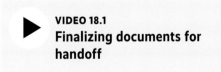

VIDEO 18.1
Finalizing documents for handoff

Packaging Files

Packaging files helps ensure efficient handoffs, by gathering all the document's elements and generating a folder containing the Illustrator file, any necessary fonts, linked assets, and a report that includes the information about the packaged files.

Package an Illustrator file

With the document open, choose **File** > **Package** to open the **Package** dialog box. Do any of the following and then click **Package** (Figure 18.8):

- Choose a location to save the packaged folder.

- Assign a name for the packaged folder.

- Select **Copy Links** to include a copy of any linked files.

- Select **Collect Links** in a **Separate Folder** to create a **Links** folder and place any linked assets within it.

- Select **Relink Linked files to the Document** to modify the link path to the package folder.

- Select **Copy Fonts (Except Adobe Fonts and non-Adobe CJK fonts)** to include all needed locally installed fonts used in the document.

- Select **Create Report** to generate and include a summary text file containing information about any relevant spot colors, fonts, and linked or embedded assets.

Exporting Files Using Export As

Illustrator provides a number of options and format types for exporting files and selected artwork.

Export a file

Do the following (**Figure 18.9**):

1. Choose **File** > **Export** > **Export As** to open the dialog box.

2. Select a location and apply a name for the file.

3. Choose a **Format** (macOS) or **Type** (Windows) for the file.

4. (Optional) if the file has multiple artboards, specify how to export them.

5. Click **Export** (macOS) or **Save** (Windows).

TIP Multiple artboards can be exported using JPEG, PNG, TIFF, and PSD formats only.

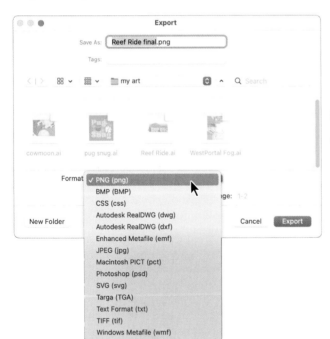

FIGURE 18.9 Exporting an Illustrator file using a PNG raster format in the Export dialog box

Customize JPEG export options

JPEG (*Joint Photographic Experts Group*) is the commonly used format for photographs. They retain most of the image's color information but compress the file size by removing certain data.

With **JPEG (jpg)** selected as the file format or type in the **Export** dialog box, do any of the following (**Figure 18.10**):

- Choose the appropriate **Color Model** from the menu.

- Adjust the **Quality** using the slider or menu to determine the quality and size of the file.

- Choose a **Compression Method** from the menu:

  Baseline (Standard) is most commonly used and recognized by most web browsers.

  Baseline Optimized generates a file optimized for color with a slightly smaller file size.

  Progressive displays a series of increasingly detailed scans (you specify how many) as the image downloads on web applications.

- Choose a **Resolution** setting from the menu. Choosing **Custom** lets you specify the value.

- Choose an **Anti-Aliasing** option to adjust the appearance of jagged edges in the rasterized artwork.

- If the file has an associated imagemap in the **Attributes** panel, select **Image-map** if you want to generate code for it.

- Select **Embed ICC Profile** to include the active ICC profile with the JPEG file.

TIP Baseline Optimized and Progressive JPEG images are not supported by all web browsers.

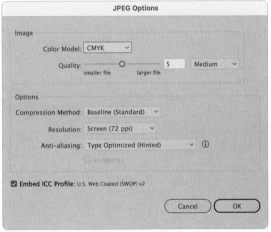

FIGURE 18.10 The JPEG Options dialog box for exporting files

Customize PNG export options

PNG (*Portable Network Graphics*) is an image file format that supports lossless data compression and transparency.

With **PNG (png)** selected as the file format in the **Export** dialog box, do any of the following (**Figure 18.11**):

- Choose a **Resolution** option from the menu. Higher resolutions create better image quality but also result in larger file sizes.

  TIP Some applications will open PNG files using 72 ppi and ignore the specified setting.

- Choose an **Anti-Aliasing** option to adjust the appearance of jagged edges in the rasterized artwork.

- Select **Interlaced** if you want to display a low-resolution version of the image as the file downloads in a browser.

- Choose a **Background Color** from the menu to specify what color is used for filling transparency. If you want to preserve the transparency, choose **Transparent**.

FIGURE 18.11 The PNG Options dialog box

Customize TIFF export options

TIFF (*Tagged-Image File Format*) is a flexible raster format that is widely supported by most applications and platforms.

With **TIFF (tif)** selected as the file format in the **Export** dialog box, do any of the following (**Figure 18.12**):

- Choose an appropriate **Color Model** from the menu for the exported file.

- Choose a **Resolution** option from the menu. Higher resolutions create better image quality but also result in larger file sizes.

- Choose an **Anti-Aliasing** option to adjust the appearance of jagged edges in the rasterized artwork.

- Select LZW compression to apply a lossless compression method that results in a smaller file size and does not discard any of the image's detail.

- Select **Embed ICC Profile** to include the ICC profile with the TIFF file. This applies to all formats that can save a color profile.

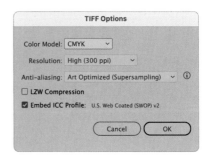

FIGURE 18.12 The TIFF Options dialog box

Customize Photoshop export options

Exporting as a Photoshop document allows you to preserve layers and editable text features, unless the artwork contains data that cannot be converted to that format.

With **Photoshop (psd)** selected as the file format in the **Export** dialog box, do any of the following (**Figure 18.13**):

- Choose an appropriate **Color Model** from the menu for the exported file. (If including layers, do not change the color mode.)

- Choose a **Resolution** option from the menu. Higher resolutions create better image quality but also result in larger file sizes.

- Select **Flat Image** if you want to merge all layers while preserving the visual appearance of the image.

- Select **Write Layers** to export layers, groups, and compound shapes as separate editable layers in Photoshop.

- Select **Preserve Text Editability** to include point type elements.

- Select **Maximum Editability** to convert all top-level sublayers to individual layers in Photoshop if this process does not compromise the artwork appearance.

- Choose an **Anti-Aliasing** option to adjust the appearance of jagged edges in the rasterized artwork.

- Select **Embed ICC Profile** to include the ICC profile with the Photoshop file. This applies to all formats that can save a color profile.

FIGURE 18.13 The Photoshop Export Options dialog box

Other formats for exporting files

AutoCAD Drawing (DWG) and **AutoCAD Interchange File (DXF)** are the standard file formats for saving vector graphics created with the AutoCAD application.

BMP is a standard image format for Windows. Illustrator lets you specify the color model, resolution, and anti-alias rasterizing settings for rasterizing the artwork, as well as a format (Windows or OS/2) and a bit depth.

Enhanced Metafile (EMF) is a common Windows format for exporting vector graphics data. However, some vector data may be rasterized.

Targa (TGA) is used for the animation and video gaming industries. Illustrator lets you specify the color model, resolution, and anti-alias rasterizing settings for rasterizing the artwork, as well as a bit depth.

Text Format (TXT) in Illustrator lets you export the file's text elements.

Windows Metafile (WMF) is a format for intermediary exchange between 16-bit Windows applications. Its vector graphics support is limited, and therefore it is recommended to use EMF format instead whenever possible.

Exporting with Export for Screens

The **Export for Screens** dialog box lets you use a single action to export multiple raster, PDF, SVG, and OBJ files with different formats and settings for use with digital devices.

Export documents and artboards using Export for Screens

Do the following (**Figure 18.14**):

1. Click **File** > **Export** > **Export for Screens**.

2. Make sure that the **Artboards** tab is active.

3. Choose which artboards you want to export by clicking the thumbnails or by selecting an option under the **Select** section:

   All selects all artboards and exports them individually.

   Range selects artboards singularly or within a range and exports them individually.

   Full Document exports all the artboards as one file.

4. Under **Export To**, choose the location folder for the exported files, how you want the exported files organized, and whether you want to view the folder after the export is completed.

5. Under **Formats**, do any of the following:

   Customize the initial export format **Scale**, **Suffix**, and **Format** options as needed.

   Add and customize additional export formats by clicking **Add Scale**.

   Click **iOS** or **Android** to set up a range of different document formats and scales that you typically need for these operating systems so you don't have to set them up individually (and repeatedly).

   Click the **Advanced Settings** button to further configure the file format options.

   Enter a **Prefix** if you want to include a description at the beginning of the generated file names.

   Click the **X** button to delete the format.

6. Click **Export Artboards** to generate the exported files.

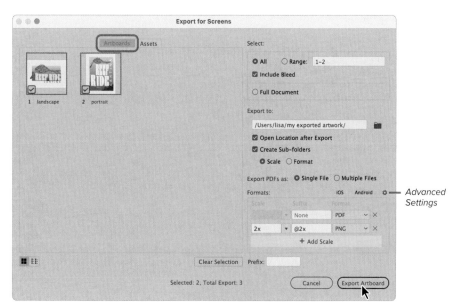

Advanced
Settings

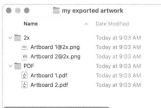

FIGURE 18.14 Exporting all of a file's artboards using PDF and PNG formats

Export assets using Export for Screens

Using the **Assets** section of the **Export for Screens** dialog box lets you export selected elements individually.

Do the following (**Figure 18.15**):

1. Select the artwork and then choose **File** > **Export** > **Export Selection**.

2. Make sure that the **Assets** tab is active.

3. Choose which assets you want to export by either clicking the thumbnails or selecting or deselecting **All Artboards**.

4. Under **Export To**, choose the location folder for the exported files, how you want the exported files organized, and whether you want to view the folder after the export is completed.

5. Under **Formats**, do any of the following:

   Customize the initial export format **Scale**, **Suffix**, and **Format** options as needed.

   Add and customize additional export formats by clicking **Add Scale**.

   Click **iOS** or **Android** to set up a preset format that you typically need for these operating systems so you don't have to set it up manually (and repeatedly).

   Click the **Advanced Settings** button to further configure file format options.

   Enter a **Prefix** if you want to include a string at the beginning of the file names generated.

   Click the **X** button to delete the format.

6. Click **Export Asset** to generate the files.

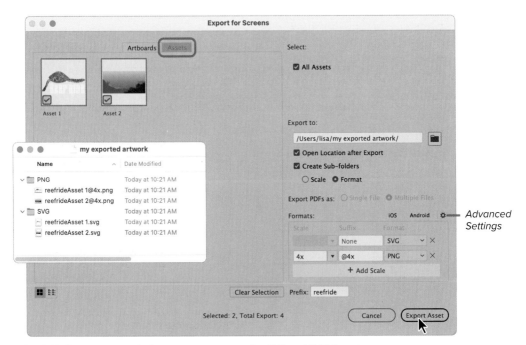

FIGURE 18.15 Exporting all of a file's selected assets using SVG and PNG formats

Using the Asset Export panel

The **Asset Export** panel (**Figure 18.16**) behaves similarly to the **Export for Screens** dialog box, with additional features for managing the collected assets.

> **TIP** To open the Asset Export panel, choose **Window > Asset Export.**

Export collected assets

Do the following:

1. Select the assets you want to export by clicking their thumbnail.

2. Customize the format options as needed.

3. (Optional) Use the panel menu (**A** in Figure 18.16) to select or deselect using submenus for the exported files.

4. Click **Export** to open the dialog box.

5. Select a folder location and then click **Choose.**

> **TIP** For more information about selecting format options, see "Exporting with Export for Screens" in this chapter.

Gather artwork as a single asset

With the artwork selected, do either of the following:

- Press **Alt/Option** while dragging the selections onto the **Asset Export** panel.
- Click the **Generate a Single Asset from Selection** button (**B** in Figure 18.16).

> **TIP** Grouped selections are treated as individual assets.

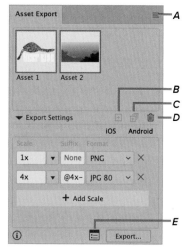

FIGURE 18.16
A. Panel menu
B. Generate Single Asset from Selection
C. Generate Multiple Assets from Selection
D. Remove Selected Asset from this Panel
E. Launch Export for Screens

Gather artwork as multiple assets

With the artwork selected, do either of the following:

- Drag the selections onto the **Asset Export** panel.
- Click the **Generate Multiple Assets from Selection** button (**C** in Figure 18.16).

Remove assets from the collection

With the asset thumbnail selected, do the following:

Click the **Remove Selected Asset from this Panel** button (**D** in Figure 18.16).

> **TIP** Removing an asset from the Asset Export panel also removes it from the Export for Screens dialog box.

Working with Save for Web

The **Save for Web** dialog box is a legacy Illustrator feature that lets you export optimized copies of your artwork using customizable presets most commonly used with web applications.

TIP Only one artboard can be exported, so be sure the one you want to use is active.

Access the Save for Web dialog box

With the file you want to export open, do the following:

- Choose **File** > **Export** > **Save for Web (Legacy)**.

Use the preview controls

In the **Save for Web** dialog box, do any of the following:

- Select the **Original** tab to display only a preview of the original artwork.

- Select the **Optimized** tab to display only a preview of the exported file.

- Select the **2-Up** tab to display both the original and exported file previews for comparison.

Customize GIF optimization settings

Under **Preset**, choose a **GIF** option from the **Name** menu or from the file type menu and then do any of the following (**Figure 18.17**):

- Adjust the **Lossy** amount to determine how much the file size is reduced using further compression by altering the dithered areas.

- Adjust the number of **Colors**.

- Choose a **Color Reduction Algorithm** from the menu to the left of the **Colors** setting.

- Adjust the level of **Dither** to determine how two adjacent colors behave to create a third.

- Choose a **Dither Algorithm** from the menu to the left of the **Dither** setting.

- Select whether **Transparency** is on or off and choose a **Matte** option.

- Under **Image Size**, adjust the dimensions and scale and choose a method of anti-aliasing.

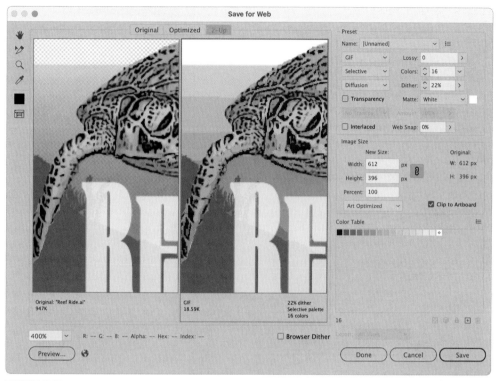

FIGURE 18.17 The Save for Web dialog box displaying the original file and a 16-color GIF with reduced dithering and no transparency previewed

Customize JPEG optimization settings

Under **Preset**, choose a **JPEG** option from the **Name** menu or from the file type menu and then do any of the following (**Figure 18.18**):

- Adjust the compression quality by entering a new value or choosing a different setting from the menu to the right of the **Quality** menu.

- Adjust the **Blur** value.

- Select or deselect **Progressive** and/or **ICC Profile**, as needed.

- Choose the **Matte** option to determine the background color.

- Under **Image Size**, adjust the dimensions and scale and choose a method of anti-aliasing.

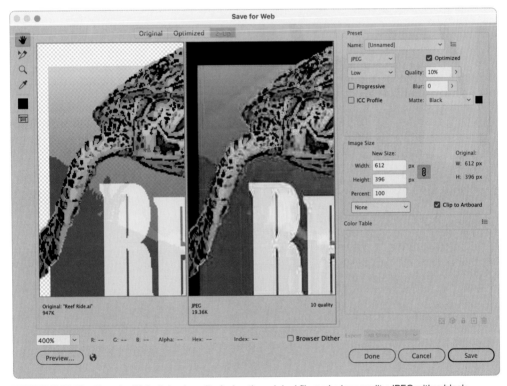

FIGURE 18.18 The Save for Web dialog box displaying the original file and a low-quality JPEG with a black background previewed

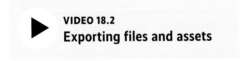

VIDEO 18.2
Exporting files and assets

Illustrator Preferences

Illustrator preferences are actually a single file that controls how the application behaves and appears. Most of these settings are accessible in the **Preferences** dialog box.

Access the Preferences dialog box

Do any of the following:

- In Windows, choose **Edit** > **Preferences** > [*preference set name*] (Windows).
- On macOS, choose **Illustrator** > **Preferences** > [*preference set name*].
- With no objects selected, click the **Preferences** button on the **Control** panel.

General

Keyboard Increment determines the distance that a selected object moves when pressing the arrow keys.

Constrain Angle determines the angle for the x and y axes (from −360° to 360°). This value affects creating new objects, transformation edits, measurements, Smart Guides, and the grid.

Corner Radius determines the initial curvature degree for object corners drawn with the **Rounded Rectangle** tool. (This value can also be set in the **Rounded Rectangle** dialog box.)

Disable Auto Add/Delete determines the **Pen** tool's ability to convert to the **Add Anchor Point** tool temporarily when positioned over a path segment on a selected path, or to the **Delete Anchor Point** tool when positioned over an anchor point on a selected path. (Pressing **Shift** enables or disables this option.)

Use Precise Cursors displays the drawing and editing tool cursors as crosshairs rather than tool icons. (Pressing **Cap Lock** turns this option on temporarily if it is disabled.)

Show Tool Tips displays a description of features such as tools, swatches, and buttons when the cursor hovers over them.

Show/Hide Rulers enables the rulers to display or hide in every document by default.

Anti-Aliased Artwork displays smoother vector edges onscreen. This does not affect printing outputs.

Select Same Tint % enables the **Select** > **Same** > **Fill Color** and **Stroke Color** commands to select only objects with the exact tint percentage as well as the same spot or global color.

Use Legacy "New File" Interface disables the latest New Document interface and reverts to the interface provided in versions CC 2015.3 and earlier.

Use Preview Bounds includes an object's stroke weight and any applied effects as part of the object's height and width dimensions. This affects transformation edits and alignment commands and the size of the bounding box.

Display Print Size at 100% Zoom matches the artwork print size to the monitor's display, regardless of the resolution. We recommend disabling this option for digital projects.

Append [Converted] upon Opening Legacy Files directs Illustrator files created in version 10 or earlier to be saved using the **Save As** dialog box and converts them into a Creative Suite (CS) or Creative Cloud (CC) version of the application. It also adds the label "[Converted]" to the file name, which prevents the file from being overwritten. If this option is not active, Illustrator saves the file in its original version, which may impact its type, editing, and appearance features.

Show System Compatibility Issue at Startup checks your computer's compatibility with the application to help determine if you need to update any drivers.

Double Click to Isolate enables double-clicking an object or group to activate Isolation Mode.

Use Japanese Crop Marks directs Illustrator to add Japanese-style crop marks when outputting color separations.

Transform Pattern Tiles enables transformation tool actions applied to an object that contains a pattern to apply the transformation to the pattern as well.

Transform Each, and **Transform Effect** dialog boxes, the **Transform** panel menu, and individual transformation tools.)

Scale Corners enables the rounded, inverted rounded, and chamfered corners of live shape objects to be scaled proportionally with the object. (This option can also be turned on or off in the **Scale** dialog box.)

Scale Strokes & Effects enables an object's stroke weight and effect settings to be scaled proportionally with the object. (This option can also be turned on or off in the **Scale** dialog box.)

Enable Content Aware Defaults allows Illustrator to use content awareness calculations when cropping images, performing puppet warp actions, and applying freeform gradients.

Honor Scale on PDF Import maintains the scale of a PDF document when opening it in Illustrator.

Zoom with Mouse Wheel lets you increase or decrease the magnification by scrolling the mouse wheel, or pan by click+dragging it.

Trackpad Gesture to Rotate View lets you rotate the canvas view using the two-finger gesture on the trackpad.

Reset All Warning Dialogs reactivates all warning boxes for which you previously selected **Don't Show Again**.

Reset Preferences restores all Illustrator preferences after you click the button, and then exit and relaunch the application.

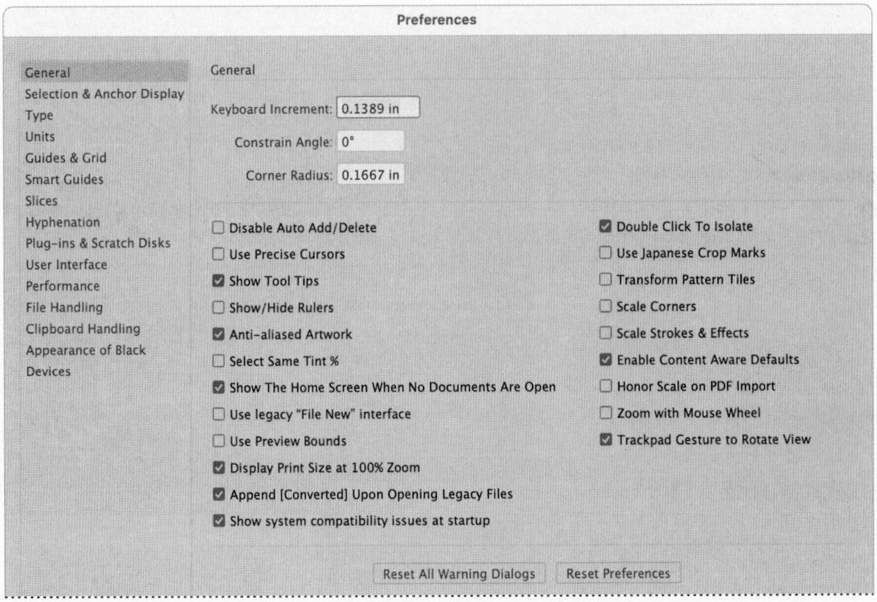

FIGURE A.1 General settings in the Preferences dialog box

Selection & Anchor Display

Selection

- **Tolerance** specifies an anchor point's selection range using the Direct Selection tool.

- **Select and Unlock Objects on Canvas** lets you unlock individual objects by clicking the assigned lock icon, rather than all locked elements using the **Object > Unlock All** command.

- **Show Anchor Points in Selection and Shape Tools** displays an object's anchor points when it is selected and a selection or shape tool is active.

- **Constrain Path Dragging on Segment Reshape** constrains the handles of a segment in a perpendicular direction when dragging the curve segment using the Anchor Point tool.

- **Move Locked and Hidden Artwork with Artboard** directs Illustrator to include locked and hidden elements on an artboard when it is repositioned or duplicated.

- **Snap to Point** snaps the cursor to a nearby anchor point or guide as you drag, draw, or scale an object within the specified pixel distance. (This option can be turned on or off using the View menu.)

- **Object Selection by Path Only** disables the ability to click an objects's fill to select it and requires clicking an object's path or anchor point using the Selection or Direct Selection tool.

- **Command/Ctrl Click to Select Objects Behind** lets you use Command/Ctrl when clicking to select objects below the currently selected object successively.

- **Zoom to Selection** focuses magnification actions on selected elements rather than the center of the screen or designated area.

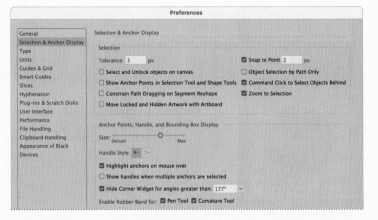

FIGURE A.2 Selection & Anchor Display settings in the Preferences dialog box

Anchor Points, Handle, and Bounding Box Display

- **Size** lets you adjust the size of the anchor points, handles, and bounding boxes.

- **Handle Style** lets you choose solid or empty circles for the direction points.

- **Highlight Anchors on Mouse Over** enlarges anchor points temporarily when the Direct Selection tool is positioned over them.

- **Show Handles when Multiple Anchors are Selected** displays all the anchor point handles for selected objects.

- **Hide Corner Widget for Angles Greater Than** lets you set the angle limit for hiding the corner widget.

- **Enable Rubber Band for** options display an interactive preview of the next curve segment for the **Pen** and/or **Curvature** tools.

Type

Size/Leading, Tracking, and Baseline Shift determine the increments selected text is modified for these values when using keyboard shortcuts to change them.

Language Options let you display East Asian (Chinese, Japanese, and Korean) or Indic (Bengali, Gujarati, Hindi, Kannada, Malayalam, Marathi, Oriya, Punjabi, Tamil, and Telugu) language characters in the **Character**, **Paragraph**, and **Open Type** panels, and in the **Type** menu.

Type Object Selection by Path Only requires clicking precisely on the type baseline when using a selection tool to select the text.

FIGURE A.3 Type settings in the Preferences dialog box

Show Font Names in English displays Chinese, Japanese, and Korean fonts using Latin characters in the Font menus instead of their native ones.

Auto Size New Area Type resizes single-column text frames to accommodate overflow text automatically.

Enable In-Menu Font Previews lets you preview the fonts in the **Character** menu for a sample of selected text before applying it.

Number of Recent Fonts determines the maximum number of recently chosen fonts (up to 15) that displays at the top of the **Font** menu and in the **Type** > **Recent Fonts** submenu.

Enable Japanese Font Preview in 'Find More' displays a simulation of the native characters for Japanese fonts when using the **Find More** tab of the **Character** menu.

Enable Missing Glyph Protection preserves glyphs placed with a non-Roman font that are converted to a Roman font.

Highlight Substituted Fonts identifies missing fonts substituted with a replacement font by highlighting them in pink for easy identification.

Fill New Type Objects with Placeholder Text automatically flows "lorem ipsum" text into newly created blank text frames.

Show Character Alternatives displays glyph alternative options in a context menu on-canvas for single selected characters.

Units

General determines the unit of measure in the currently active document for entry fields for most panels and dialog boxes and for document window rulers.

Stroke determines the unit of measure for the **Stroke** panel and the **Stroke Weight** field in the **Control** and **Appearance** panels.

Type determines the unit of measure for the **Character** and **Paragraph** panels.

East Asian Type is available only if the **Show East Asian Options** is selected in the **Type** section of the **Preferences** dialog box. The selected unit of measure applies to East Asian type only.

Numbers Without Units Are Points is available if **Picas** is the selected **General** unit of measurement. If this option is selected, values entered (excluding type and stroke fields) will be entered as points.

Identify Objects By directs Illustrator to verify that the object names in the **Layers** panel match XML specifications. (XML is an advanced feature not covered in this book.)

FIGURE A.4 Units settings in the Preferences dialog box

Guides & Grid

Guides

- **Color** lets you set the color for ruler guides by choosing a color from the menu, choosing **Custom**, or clicking the color box to open the **Colors** dialog box.

- **Style** determines whether the ruler guides display as lines or dots.

Grid

- **Color** lets you set the color for the grid by choosing a color from the menu, choosing **Custom**, or clicking the color box to open the **Colors** dialog box.

- **Style** determines whether the grid displays as lines or dots.

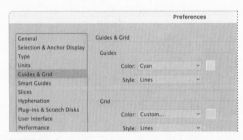

FIGURE A.5 Guides & Grid settings in the Preferences dialog box

Smart Guides

Object Guides lets you set the color for the Smart Guides by choosing a color from the menu, choosing **Custom**, or clicking the color box to open the **Colors** dialog box.

Alignment Guides displays straight line guides that indicate when an object or artboard's edge or center is aligned with another object's edge or center when manually creating, transforming, or repositioning the object.

Object Highlighting displays an outline of the original object when the cursor passes over them.

Transform Tools displays angled line guides when using the Scale, Rotate, Reflect, or Shear tools to transform an object.

Construction Guides displays angled line guides as you manually draw or transform an object, when the cursor passes over the anchor point of another object. The angles can be set using the menu or entering them in the fields.

Snapping Tolerance sets the distance (up to 10 points) within which the side, corner, or center point of an object you are creating, modifying, or repositioning needs to be from the side or center point of another object in order to snap to it.

Glyph Guides lets you snap selections precisely to text elements (x-height, baseline, or glyph boundary). The Glyph Guide options let you set the color for the glyph guides by choosing a color from the menu, choosing **Custom**, or clicking the color box to open the **Colors** dialog box.

Anchor/Path Labels displays a description (Path, Anchor, or Center) when the cursor passes over that part of an object or when paths intersect during construction.

Measurement Labels displays dimensions and x/y coordinates of objects as you manually draw, transform, or reposition them.

Spacing Guides displays line guides for positioning objects using equal distances.

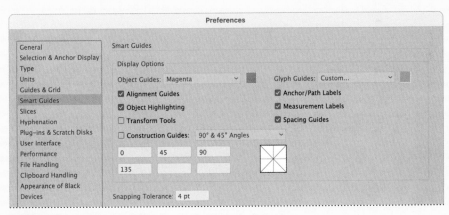

FIGURE A.6 Smart Guides settings in the Preferences dialog box

Slices

Slices are elements created for web output that correspond to a web page's table cells. They are not covered in this book.

Show Slice Numbers displays the slice number onscreen.

Line Color lets you set the color for slice boundary lines and slice number boxes by choosing a color from the menu, choosing **Custom**, or clicking the color box to open the **Colors** dialog box.

FIGURE A.7 Slices settings in the Preferences dialog box

Hyphenation

Default Language determines the dictionary language applied to characters for Illustrator to reference when hyphenating words in the current document.

Exceptions are words that you do not want Illustrator to hyphenate. They can be added by entering them in the **New Entry** field and then clicking **Add**. To remove an exception select it and click **Delete**.

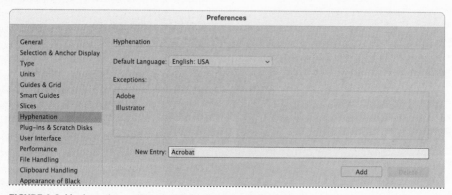

FIGURE A.8 Hyphenation settings in the Preferences dialog box

Plug-ins & Scratch Disks

Additional Plug-ins Folder lets you choose additional external plug-ins located in different locations than the ones provided with the Illustrator application.

Scratch Disks lets you choose a Primary and optional Secondary hard disk for the scratch disk that Illustrator uses as virtual memory when the available RAM is insufficient for processing.

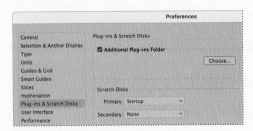

FIGURE A.9 Plug-ins & Scratch Disks settings in the Preferences dialog box

User Interface

Brightness presets adjust the workspace lightness.

Canvas Color lets you choose to match the background brightness for the document canvas (the area behind the artboard) with the workspace or choose white.

Auto-Collapse Iconic Panels closes expanded panels that were opened by clicking its thumbnail after clicking away from it.

Open Documents as Tabs docks multiple open documents using tabs in the application frame.

Large Tabs displays taller document tabs.

UI Scaling lets you adjust the application's user interface scale to best accommodate your monitor resolution.

Scale Cursor Proportionately resizes the cursor relative to the UI scale.

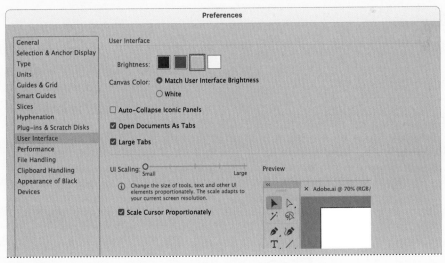

FIGURE A.10 User Interface settings in the Preferences dialog box

Performance

GPU Performance refers to the graphics processing unit that allows accelerated execution of commands for displaying and manipulating artwork, allowing Illustrator to run more efficiently.

- **Animated Zoom** lets zoom actions perform in a smooth and animated manner by dragging the **Zoom** tool left or right.

GPU Details provides information about the device and memory associated with your device.

Undo Counts determines how many **Edit** > **Undo** commands can be executed.

Real-Time Drawing and Editing displays the live appearance of selected objects as you manipulate them.

FIGURE A.11 Performance settings in the Preferences dialog box

File Handling

File Save Options

- **Automatically Save Recovery Data Every** enables Illustrator to save a recovery version of the file that it provides if your system crashes. The menu lets you determine how often the autosave occurs.

- **Folder Choose** lets you select a different folder for saving recovery files

- **Turn off Data Recovery for Complex Documents** allows Illustrator to pause backing up complex files if it impacts performance.

- **Save in Background** and **Export in Background** lets you continue to work in a document while Illustrator saves or exports it.

- **Automatically Save Cloud Documents Every** enables autosave of cloud documents. The menu lets you determine how often this occurs.

Files

- **Number of Recent Files to Display** determines how many files are displayed under **File > Open Recent Files**.

- **Optimize File Open and Save Time on Slow Network** reduces the amount of time needed to save a file to a network.

- **Use Low Resolution Proxy for Linked EPS** displays placed linked EPS images at low resolution to enhance performance.

- **Display Bitmaps as Anti-Aliased Images in Pixel Preview** softens 1-bit raster image edges when **View > Pixel Preview** is active.

- **Update Links** determines how modified linked files are updated.

- **Use System Defaults for 'Edit Original'** lets you choose which application is used when clicking **Edit Original** for placed images.

Fonts

- **Auto-activate Adobe Fonts** lets Illustrator automatically replace missing fonts using Adobe Fonts.

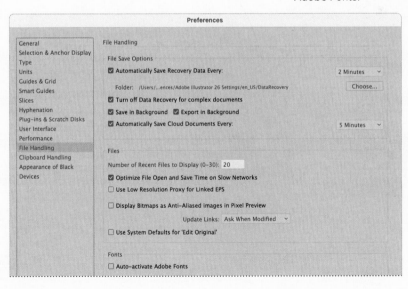

FIGURE A.12 File Handling settings in the Preferences dialog box

Clipboard Handling

On Copy's Include SVG Code option lets a copied object's SVG code be included on the Clipboard for pasting.

On Quit options determine whether Illustrator copies artwork using AICB (Adobe Illustrator Clipboard) and/or PDF file formats for pasting files.

- PDF (Portable Document Format) supports native transparency and pastes artwork as graphics instead of editable paths.

- AICB (Adobe Illustrator Clipboard Document) is a legacy Postscript format that does not support transparency. **Preserve Paths** retains the vector paths. **Preserve Appearance and Overprints** maintains appearances such as effects and overprint elements as separate objects.

Paste Text without Formatting removes inherited character and paragraph type formatting from text when pasting.

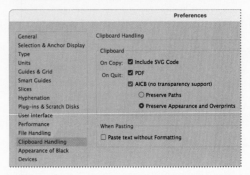

FIGURE A.13 Clipboard Handling settings in the Preferences dialog box

Appearance of Black

The **Appearance of Black** options determine how your monitor displays the color black and what black will look when output digitally (RGB) or printed (CMYK).

Onscreen

- **Display All Blacks Accurately** displays 100% black as dark gray, as when printed using CMYK.

- **Display All Blacks as Rich Black** displays all blacks as rich black regardless of the assigned CMYK values.

Printing / Exporting

- **Output All Blacks Accurately** displays all blacks as dark gray when viewed or printed using RGB devices.

- **Display All Blacks as Rich Black** displays all blacks as rich black regardless of the assigned RGB values.

Descriptions provides explanations of the **On Screen** and **Printing / Exporting** menu options when you hover the cursor over them.

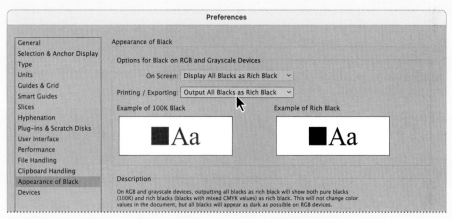

FIGURE A.14 Appearance of Black settings in the Preferences dialog box

Devices

Enable Wacom allows Illustrator to recognize Wacom tablets and styluses.

If Illustrator crashes due to an issue with a Wacom device, this option is automatically disabled, requiring you to reselect it the next time you start the application.

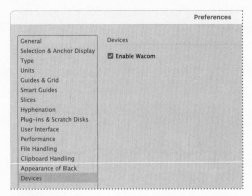

FIGURE A.15 Devices settings in the Preferences dialog box

B

Keyboard Shortcuts

Illustrator provides a vast array of keyboard shortcuts for tools and commands that help with productivity as you become more familiar with the application.

For quick reference, this appendix is a compilation of some of the most commonly used keyboard shortcuts.

Dive Deeper

You can find more information about additional shortcuts in the online **Illustrator User Guide** (helpx.adobe.com), under **Default Keyboard Shortcuts**.

Keyboard shortcuts can also be customized in the **Keyboard Shortcuts** dialog box (**Edit** > **Keyboard Shortcuts**).

Common Shortcuts

Menu shortcuts are displayed to the right of the command when you choose them from the **Menu bar**.

Toolbar shortcuts are displayed to the right of the tool name and are visible when you hover over the **Hidden Tool** icon (**Figure B.1**).

TIP Command and Option are macOS. Ctrl and Alt are Windows.

File and Illustrator menus

Action	Shortcut
Open a file	Command/Ctrl+O
Create a new file	Command/Ctrl+N
Save a file	Command/Ctrl+S
Place a file	Shift+Command/Ctrl+P
Print a file	Command/Ctrl+P
Close a file	Command/Ctrl+W
Open Preferences dialog box	Command/Ctrl+K
Quit Illustrator	Command/Ctrl+Q

Select and Object menus

Action	Shortcut
Add to selection	Shift+click or drag
Subtract from selection	Shift+click or drag
Select all	Command/Ctrl+A
Deselect	Shift+Command/Ctrl+A
Lock selection	Option/Alt+Command/Ctrl+2
Unlock all	Shift+Command/Ctrl+2
Group selection	Command/Ctrl+G
Ungroup selection	Shift+Command/Ctrl+G
Move to front	Shift+Command/Ctrl+]
Bring forward	Command/Ctrl+]
Move to back	Shift+Command/Ctrl+[
Send backward	Command/Ctrl+[

Edit menu

Action	Shortcut
Undo	Command/Ctrl+Z
Redo	Shift+Command/Ctrl+Z
Copy	Command/Ctrl+C
Cut	Command/Ctrl+X
Paste	Command/Ctrl+V
Paste in Front	Command/Ctrl+F
Paste in Back	Command/Ctrl+B
Set transformation origin point	Option/Alt+click or drag
Reapply last edit action	Command/Ctrl+D
Check Spelling	Command/Ctrl+I

View menu

Action	Shortcut
Magnify at 100%	Command/Ctrl+1
Zoom In	Command/Ctrl++ (plus)
Zoom Out	Command/Ctrl+- (minus)
View in outline/ preview toggle	Command/Ctrl+Y

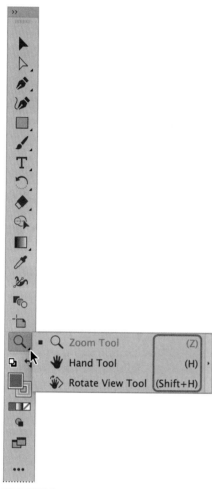

FIGURE B.1 Hovering over the Hidden Tool icon to display the Zoom tool group and the shortcuts

Toolbar

Tool		Shortcut
▶	Selection	V
▷	Direct Selection	A
✴	Magic Wand	Y
⟋	Lasso	Q
✒	Pen	P
⁺✒	Add Anchor Point	+
✒⁻	Delete Anchor Point	-
⌐	Anchor Point	Shift+C
T	Type	T
╱	Line Segment	\
▢	Rectangle	M
⬭	Ellipse	L
🖌	Paintbrush	B
✎	Pencil	N
✂	Scissors	C
↻	Rotate	R
▷◁	Reflect	O
⊡	Scale	S
▩	Gradient	G
⟍	Eyedropper	I
🔗	Blend	W
⬚	Artboard	Shift+O
✋	Hand	H
⚲	Zoom	Z
⬓	Default Fill/Stroke	D
n/a	Toggle Fill/Stroke	X
↰	Swap Fill/Stroke	Shift+X
▱	None (Fill or Stroke)	/

Index

U

Undo command
  keyboard shortcut for 288
  preferences 282
Units preferences 278
User Interface preferences 281

V

variable-width strokes 94
Vertical Type on a Path tool
  adding text with 120
Vertical Type tool
  adding text with 118
Video effects 236
View menu
  keyboard shortcuts 288
view modes
  changing 58
  Isolation 58
  Outline 58
  Preview 58

W

Wacom devices
  working with brushes 151
  enabling 286
Warning dialog boxes
  resetting 275
Warp effects 229
Warp tool
  reshaping objects with 194
white space characters
  inserting 140
Width tool
  varying stroke with 95
Window menu
  opening panels with 16
Windows Metafile (WMF)
  exporting files to 265

workspaces
  customizing 20
  Essentials Classic 3
  Manage Workspaces dialog box 21
  menu 20
  New Workspace dialog box 20
  resetting 20
  saving 20
  Switch Workspace button 20
Wrinkle tool
  reshaping objects with 200

Z

Zoom tool 53

Notes

Notes

Notes

Image Credits

WITHDRAWN

Learn the quick and easy way
VISUAL QUICKSTART GUIDES

Visual QuickStart Guides provide an easy, visual approach to learning. Your purchase of the book or eBook includes the Web Edition, a free online version of the book containing bonus video lessons that support and expand on topics covered in the book.

Concise steps and explanations let you get up and running in no time. These step-by-step tutorial and quick-reference guides are an established and trusted resource for creative people.

To see a complete list of our Visual QuickStart Guides go to:

peachpit.com/vqs

HTML and CSS:
Visual QuickStart Guide, 9th Edition
ISBN: 9780136702566

Adobe Photoshop Elements
Visual QuickStart Guide
ISBN: 9780137637010

Adobe Illustrator Visual QuickStart Guide
ISBN: 9780137597741

Adobe Photoshop Visual QuickStart Guide
ISBN: 9780137640836

Peachpit Press
www.peachpit.com